Ancient Connections in Eurasia

Words, Bones, Genes, Tools: DFG Center for Advanced Studies Series

Volume three

Ancient Connections in Eurasia

Hugo Reyes-Centeno, Katerina Harvati
Editors

Kerns Verlag Tübingen

Words, Bones, Genes, Tools: DFG Center for Advanced Studies Series

Editors of the Series: Katerina Harvati and Gerhard Jäger

Kerns Verlag
Postfach 210516, 72028 Tübingen, Germany
Fax: 49-7071-367641 Tel: 49-7071-367768
email: info@kernsverlag.com
www.kernsverlag.com

Cover Image:
Kyparissia lignite mine, Megalopolis Basin. This region is known for its rich Pleistocene fossiliferous deposits, and has yielded several Lower Paleolithic sites, including Marathousa 1, the oldest currently known radiometrically dated archaeological site in Greece. An isolated human fossil tooth from the Megalopolis Basin is analyzed in Chapter 1 of this volume.
Photo copyright K. Harvati, University of Tübingen, taken by Nicholas Thompson in the framework of the ERC CoG CROSSROADS / MEGAPAL. The MEGAPAL survey is conducted under a permit granted to the Ephoreia of Palaeoanthropology-Speleology, Greek Ministry of Culture, and the American School of Classical Studies at Athens.

Jacket Design: Hemmerich Gestaltung, Tübingen.
Printer: Sautter, Reutlingen.
Layout and Design: Kerns Verlag, Tübingen.

© 2021 Kerns Verlag.
© 2021 The name of the series belongs exclusively to the DFG Center for Advanced Studies, University of Tübingen. For permission requests to reproduce material from this publication, please contact the publisher.
Alle Rechte vorbehalten | All rights reserved.
ISBN: 978-3-935751-37-7.
Printed in Germany.

Table of Contents

Preface	*Hugo Reyes-Centeno, Katerina Harvati*	7
Chapter 1	Crown outline analysis of the hominin upper third molar from the Megalopolis Basin, Peloponnese, Greece *Carolin Röding, Julia Zastrow, Heike Scherf, Constantin Doukas and Katerina Harvati*	13
Chapter 2	Direct U-series dating of the Apidima C human remains *Katerina Harvati, Rainer Grün, Mathieu Duval, Jian-xin Zhao, Alexandros Karakostis, Vangelis Tourloukis, Vassilis Gorgoulis, Mirsini Kouloukoussa*	37
Chapter 3	Profile orientation change through time in Upper Paleolithic parietal art *Matteo Scardovelli*	57
Chapter 4	High level connections as a key component for the rapid dispersion of the Neolithic in Europe *Solange Rigaud*	73
Chapter 5	Genetic demography: What does it mean and how to interpret it, with a case study on the Neolithic transition *Michela Leonardi, Guido Barbujani and Andrea Manica*	91
Chapter 6	Statistical methods for kinship Inference amongst ancient individuals *Andaine Seguin-Orlando*	101
Chapter 7	The structure of cranial morphological variance in Asia: Implications for the study of modern human dispersion across the planet *Mark Hubbe*	129
Chapter 8	Associations between human genetic and craniometric differentiation across North Eurasia: The role of geographic scale *Andrej Evteev, Patrícia Santos, Alexandra Grosheva, Hugo Reyes-Centeno, Silvia Ghirotto*	157

Preface

This volume collects the proceedings of the fourth annual symposium of the DFG Center for Advanced Studies "Words, Bones, Genes, Tools: Tracking linguistic, cultural and biological trajectories of the human past." The symposium, titled Ancient Connections in Eurasia, explored linguistic, cultural, and biological trends in Eurasian prehistory since the Paleolithic. More than thirty scholars, representing institutions around the world, gathered in Tübingen on December 13th-16th, 2018, to present and discuss their findings in an intimate and lively context. In keeping with the Center's focus on interdisciplinary research, presentations covered a range of topics, including linguistic, archaeological, genetic and anthropological approaches.

This volume comprises eight chapters representing most of the work presented at the symposium, as well as additional contributions. These include both original research and review or perspective pieces. The volume is arranged in two broad thematic units: The first comprises contributions focusing on specific sites or time periods (Chapters 1–4); whereas the second collects contributions aiming to decipher human genetic and cranial variation to better understand evolutionary and demographic processes in the past (Chapters 5-8).

Chapter 1 (Röding et al., this volume) is a case study of an isolated human tooth proposed to be a possible Middle Pleistocene fossil from Greece. This specimen was recovered in the 1960s together with fossil fauna in the Megalopolis Basin, a region well known for its Pleistocene fossiliferous deposits, and recently yielding the oldest archaeological evidence for human presence in Greece and possibly southeast Europe (see, e.g., Harvati et al. 2018). In light of these developments, the Megalopolis tooth takes on particular importance; however, its lack of provenance, its status as an isolated upper third molar and surface abrasion of its crown make it difficult to assess. Röding and colleagues undertake the

© 2021, Kerns Verlag / https://doi.org/10.51315/9783935751377.pre
Cite this article: Reyes-Centeno, H., and K. Harvati. 2021. Preface. In *Ancient Connections in Eurasia*, ed. by H. Reyes-Centeno and K. Harvati, pp. 7-11. Tübingen: Kerns Verlag. ISBN: 978-3-935751-37-7.

difficult task of analyzing this specimen using state-of-the-art methodology of semilandmark-based comparative analysis of its crown outline. Results suggest a tentative attribution of the specimen to the Neanderthal lineage, supporting the proposed fossil status of the specimen and highlighting the Megalopolis Basin as a key area for uncovering the deep human past of Eurasia.

Chapter 2 (Harvati et al., this volume) continues with work on the human fossil record of Greece, focusing on the human burial from the Apidima site (Cave C). Apidima is one of the most important fossil-human-bearing sites of southeast Europe, yielding two Middle Pleistocene human crania from Cave A (Harvati et al. 2019), as well as a burial from Cave C. The latter has been proposed to be of possible Aurignacian age, although neither chronometric dates nor a secure association with an Upper Paleolithic context exist. Harvati and colleagues obtained direct U-series dates from the burial, as well as from two isolated human teeth representing different individuals, in order to place these remains in a secure chronological framework. Results confirm a Pleistocene geological age for the Cave C skeletons, but further analysis will be necessary to further refine their chronology.

In Chapter 3, Scardovelli (this volume) explores the diachronic variability of Upper Paleolithic cave paintings, focusing on images from twenty-six caves in France and Spain spanning the time period from the Aurignacian to the Magdalenian. Using statistical analysis of the profile orientation of more than 2000 images, Scardovelli (this volume) detects a major dichotomy between the Aurignacian and the later Upper Paleolithic cultural traditions analyzed. His findings are presented in the context of cultural, biomechanical, and neurological variables possibly affecting image orientation, and suggest a cultural shift, possibly associated with a population shift, at the end of the Aurignacian. Interestingly, these results might be consistent with a recently proposed hypothesis, developed from paleogenetic as well as craniometric data, that the Gravettian originated in Eastern Europe and spread across Western Europe in a population movement accompanied by cultural developments (Bennett et al. 2019; Mounier et al. 2020).

Chapter 4 (Rigaud, this volume) closes the first thematic unit, and also focuses on cultural remains, this time personal ornaments. The chapter reviews how animal tooth pendants and shell beads in particular can be used to infer the cultural and economic context of past societies across time and space. In particular, Rigaud (this volume) uses data from the late Mesolithic and early Neolithic of Europe to argue that personal ornaments reflect rapid, fundamental changes in the belief systems and social configurations of populations adopting agricultural economies and sedentary lifestyles. Such rapid change is hypothesized to have been facilitated by the extensive network developed between Mesolithic populations.

Chapter 5 (Leonardi et al., this volume) opens the second thematic unit with a review of the concept of genetic demography and its applica-

Fig. 1.
Symposium participants in front of the *Alte Aula* historical building of the University of Tübingen on December 14th, 2018.

tion in archaeological contexts. Leonardi et al. (this volume) concentrate on the term "effective population size," which can be calculated from genetic data and is often used in archaeological interpretations. The authors clearly explain the term and the assumptions that underlie its calculation, and elucidate the possible interpretations of observed effective population size values in the past. To further illustrate their point, Leonardi et al. (this volume) summarize their recent findings on effective population size differences across the Neolithic transition, and how they can be interpreted.

Chapter 6 (Seguin-Orlando, this volume) provides a review of cutting-edge statistical methods for inferring kinship amongst individuals in the past. The first section reviews methods and theory, while the second section concentrates on recent applications, focusing on the nuances of working with ancient DNA. While population-level paleogenomic studies have been critical in reconstructing Eurasian prehistory, Seguin-Orlando (this volume) highlights how family-level studies can greatly enrich our understanding of the human past, from addressing pivotal anthropological questions about social structure and migration to understanding local funerary practices and contextualizing associated material culture.

Chapter 7 (Hubbe, this volume) is a worldwide survey of cranial morphological variation in modern humans, assessing whether patterns in Asian populations can be explained by the continent's complex history of population movements. While several global-scale studies have shown that cranial morphological variation is structured largely as a result of the expansion of modern humans out of Africa tens of thousands of years ago, the effect of more recent population movements remains poorly understood. Using a quantitative genetics approach to quantify the apportionment of cranial morphological variance in Asia compared to the rest of the world, Hubbe (this volume) contextualizes the results against the backdrop of four primary human dispersion stages inferred from the archaeological and paleontological records. Because morphological variation in some Asian populations depart from expectations, Hubbe cautions on the use of craniometric data alone for inferring past population movements, particularly at local rather than global scales.

In the concluding contribution, Chapter 8 (Etveev et al., this volume), the authors continue in a similar vein, exploring the effects of geographical scale, as well as cranial region and type of genetic dataset, on the correlation between craniometric, genetic and geographic distances in a large original craniometric dataset from Eurasia. Such correlations have been found in previous research (e.g., Harvati and Weaver 2006; Reyes Centeno et al. 2017) and are commonly interpreted to support the utility of cranial anatomy in reconstructing population history. The work of Etveev et al. (this volume) is an important methodological contribution, which clarifies the importance of geographic scale, as correlations between craniometric and genetic data decrease with smaller geographic distances, greatly reducing the utility of this approach to address questions of local, rather than global, nature.

In closing, we are thankful to all the participants of the symposium, as well as all the authors who contributed their manuscripts for inclusion in this volume. We are also grateful to George Perry, Miriam Haidle, Christian Benz and Gerhard Jäger for chairing sessions, and, as ever, to Monika Doll, whose dedication, hard work and organizational skills were crucial to the success of the symposium. We also thank all the University of Tübingen students and members of the DFG Centre for Advanced Studies who helped with the organization and during the symposium: Abel Bosman, Alessio Maiello, Saetbyul Lee, Patrícia Silva Santos, Caro Röding, Julia Zastrow, David Naumann and Gabriele Russo; and the colleagues who agreed to review the manuscripts published here. We remain deeply grateful to the University of Tübingen President, Dr. Bernd Engler, and Dr. Peter Grathwohl, Vice President for Research, for their continuing support. Funding for the conference was provided by the *Deutsche Forschungsgemeinschaft*, in the framework of the *DFG Kollegforschergruppe* "Words, Bones, Genes, Tools" (DFG FOR 2237). Last but not least, we thank our families for their patience and support.

Hugo Reyes-Centeno, Katerina Harvati
December 2020

REFERENCES

Bennett, E. A., S. Prat, S. Péan, L. Crépin, A. Yanevich, S. Puaud, T. Grange, and E.-M. Geigl. 2019. The origin of the Gravettians: Genomic evidence from a 36,000-year-old Eastern European. bioRxiv https://doi.org/10.1101/685404

Evteev, A., P. Santos, A. Grosheva, H. Reyes-Centeno, and S. Ghirotto. Associations between human genetic and craniometric differentiation across North Eurasia: The role of geographic scale. In *Ancient Connections in Eurasia*, ed. by H. Reyes-Centeno and K. Harvati. Tübingen: Kerns Verlag.

Harvati, K., and T. Weaver. 2006. Human cranial anatomy and the differential preservation of population history and climate signatures. *Anatomical Record* 288A: 1225–1233

Harvati, K., C. Röding, A. M. Bosman, F. A. Karakostis, R. Grün, C. Stringer, P. Karkanas, N. C. Thompson, V. Koutoulidis, L. A. Moulopoulos, V. G. Gorgoulis, and M. Kouloukoussa. 2019. Apidima Cave fossils provide earliest evidence of *Homo sapiens* in Eurasia. *Nature* 571: 500–504.

Harvati, K., R. Grün, M. Duval, J. Zhao, A. Karakostis, V. Tourloukis, V. Gorgoulis, and M. Kouloukoussa. 2021. Direct U-series dating of the Apidima C human remains. In *Ancient Connections in Eurasia*, ed. by H. Reyes-Centeno and K. Harvati. Tübingen: Kerns Verlag.

Hubbe, M. 2021. The structure of cranial morphological variance in Asia: Implications for the study of modern human dispersion across the planet. In *Ancient Connections in Eurasia*, ed. by H. Reyes-Centeno and K. Harvati. Tübingen: Kerns Verlag.

Leonardi, M., G. Barbujani, and A. Manica. 2021. Genetic demography: What does it mean and how to interpret it, with a case study on the Neolithic transition. In *Ancient Connections in Eurasia*, ed. by H. Reyes-Centeno and K. Harvati. Tübingen: Kerns Verlag.

Mounier, A., Y. Heuzé, M. Samsel, S. Vasilyev, L. Klaric, and S. Villotte. 2020. Gravettian cranial morphology and human group affinities during the European Upper Palaeolithic. *Scientific Reports* 10: 21931.

Reyes-Centeno, H., S. Ghirotto, and K. Harvati. 2017. Genomic validation of the differential preservation of population history in the human cranium. *American Journal of Physical Anthropology* 162: 170–179

Rigaud, S. 2021. High level connections as a key component for the rapid dispersion of the Neolithic in Europe. In *Ancient Connections in Eurasia*, ed. by H. Reyes-Centeno and K. Harvati. Tübingen: Kerns Verlag.

Röding, C., J. Zastrow, H. Scherf, C. Doukas, and K. Harvati. 2021. Crown outline analysis of the hominin upper third molar from the Megalopolis Basin, Peloponnese, Greece. In *Ancient Connections in Eurasia*, ed. by H. Reyes-Centeno and K. Harvati. Tübingen: Kerns Verlag.

Sardovelli, M. 2021. Profile orientation change through time in Upper Paleolithic parietal art. In *Ancient Connections in Eurasia*, ed. by H. Reyes-Centeno and K. Harvati. Tübingen: Kerns Verlag.

Chapter 1

Crown outline analysis of the hominin upper third molar from the Megalopolis Basin, Peloponnese, Greece

Carolin Röding[1], Julia Zastrow[2], Heike Scherf[1], Constantin Doukas[3] and Katerina Harvati[1,2,4]

Abstract

The left upper third molar from the Megalopolis Basin is enigmatic due to its problematic preservation and context. The Megalopolis molar is the only possible human fossil known to date from the Megalopolis Basin. It was found on the surface during geological surveys in 1962-63. Based on the faunal assemblage collected during the same survey, it was proposed to be of Middle Pleistocene age and possibly one of the oldest human fossils in Europe. However, its actual geological age is unknown. In the past, dental crown outline analysis has been successfully used to differentiate between hominin species and populations. We applied the method to upper third molars, attempting to shed light on the affinities of the Megalopolis specimen. Principal component analysis (PCA) of the crown outline shape grouped the Megalopolis molar with our *Homo sapiens* sample; however, the PCA in form space, including shape plus size, as well as Procrustes distances based on overall shape, grouped it with our Neanderthal comparative sample. We conclude that its most likely identification is as a member of the Neanderthal lineage. However, we urge further analyses with an increased fossil comparative sample to include representatives of *Homo heidelbergensis*, which is underrepresented in our study. The Megalopolis molar contributes to the scarce Pleistocene human fossil record of Greece and highlights the potential of the Megalopolis Basin for yielding further paleoanthropological finds.

---

1 Paleoanthropology, Senckenberg Centre for Human Evolution and Palaeoenvironment, Eberhard Karls University of Tübingen, Germany.
2 DFG Centre of Advanced Studies 'Words, Bones, Genes, Tools', Eberhard Karls University of Tübingen, Germany.
3 Faculty of Geology and Geoenvironment, National and Kapodistrian University of Athens, Greece.
4 Museum of Anthropology, Medical School, National and Kapodistrian University of Athens, Greece.

© 2021, Kerns Verlag / https://doi.org/10.51315/9783935751377.001
Cite this article: Röding, C., J. Zastrow, H. Scherf, C. Doukas, and K. Harvati. 2021. Crown outline analysis of the hominin upper third molar from the Megalopolis Basin, Peloponnese, Greece. In *Ancient Connections in Eurasia*, ed. by H. Reyes-Centeno and K. Harvati, pp. 13-36. Tübingen: Kerns Verlag. ISBN: 978-3-935751-37-7.

INTRODUCTION

The Megalopolis Basin, Peloponnese, Greece, is well-known for its fossil fauna (e.g., Skoufos 1905; Melentis 1961; Sickenberg 1976; Athanassiou 2018; Athanassiou et al. 2018) and more recently for its Middle Pleistocene archaeological sites (e.g., Panagopoulou et al. 2015; Giusti et al. 2018; Thompson et al. 2018; Konidaris et al. 2019). The most important of these, Marathousa 1, has yielded a stratified lithic as well as faunal assemblage including elephant remains showing signs of butchery (e.g., Tourloukis et al. 2018a; Konidaris et al. 2018). The site has been dated to 400-500 ka (Blackwell et al. 2018; Jacobs et al. 2018), testifying to an early human presence in the region. In contrast to these recently discovered sites, many of the earlier paleontological finds from the Megalopolis Basin are non-stratified surface finds. Surface finds can be transported and can originate from varying exposed surfaces in the proximity of the find spot, which complicates their dating (Wandsnider 2004). In the case of the Megalopolis area, exposed surfaces span a wide geological age range (Siavalas et al. 2009; Vinken 1965). In 1962-63 an isolated human tooth was found on the surface in the basin and recovered together with Pleistocene fossil faunal remains (Sickenberg 1976; Marinos 1975). This putative fossil human specimen is a left upper third molar, hereafter referred to as the Megalopolis molar (Fig. 1). Its geological age and species attribution are unknown because of its problematic context as surface find, as well as its state of preservation.

The Megalopolis molar was first described during the analysis of the faunal remains collected at the same time (Marinos 1975). It was proposed that the Megalopolis molar has a similar age as the fauna. The faunal assemblage was assigned to the "Biharium" (Sickenberg 1976), which roughly translates to the lower half of the Middle Pleistocene and the Early Pleistocene (Koenigswald and Heinrich 2007). If Sickenberg's assessment was correct, the Megalopolis molar would be one of the oldest hominin fossils known in Europe at the time of its discovery. In some cases, ESR and U-series dating enable direct dating of teeth (e.g., Duval et al. 2012). In the case of the Megalopolis molar, direct dating has not been attempted due to previous chemical treatment (Xirotiris et al. 1979) but also because of the destructive nature of these dating methods. Because of its status as a surface find without datable surrounding context, its fossil status is uncertain, as it could potentially derive from a recent, modern human skeleton (Marinos 1975).

Xirotiris et al. (1979) analyzed the enamel prism structure via scanning electron microscopy (SEM) with the aim to classify the Megalopolis molar. For this purpose, a part of the crown was cleaned with acid to remove the enamel surface layer, which does not show a prism structure (Xirotiris et al. 1979). The SEM method usually requires a thin gold or platinum coating of the sample to improve contrast and the signal-to-noise ratio (Carter and Shieh 2015: 117–144). Remnants of the gold coating are still visible on the fossil (Fig. 1 a-e). It is assumed that the gold

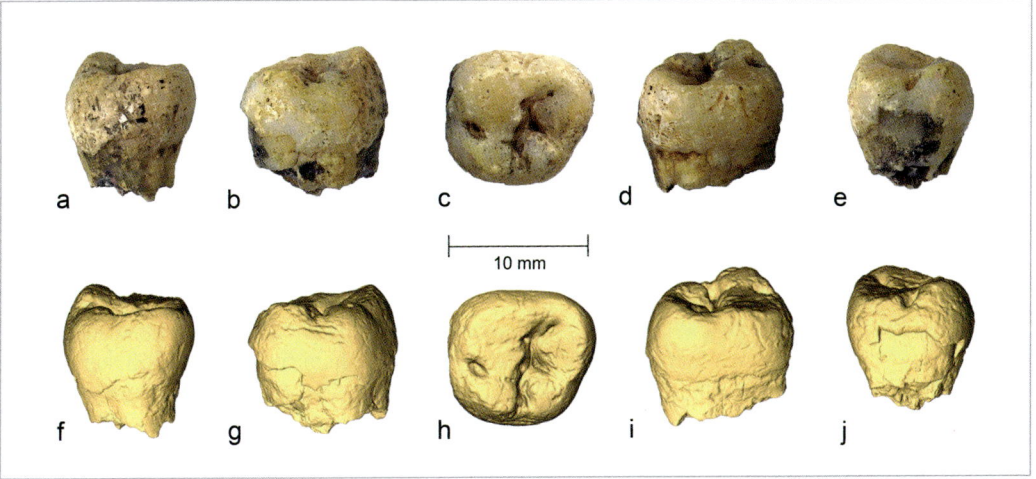

Fig. 1.
Left upper third molar from the Megalopolis Basin. Shown as pictures of the original (a-e) and surface model from a CT scan (f-j). Displayed in buccal view (a,f), distal view (b,g), occlusal view with the mesial side being oriented towards the top (c,h), mesial view (d,i) and lingual view (e,j).

coating was at least partly removed by acid due to the concomitant removal of labeling on the crown (cf. Fig. 1, Xirotiris et al. 1979: 118), which resulted in the obliteration of its crown features. The authors of that study did not reach a species attribution due to the limited comparative sample of fossil human enamel prism structures and an overlap in the linear crown measurements of the Megalopolis molar with several *Homo* species (Xirotiris et al. 1979).

The preservation of the Megalopolis molar is problematic for most standard methods used to assess external morphology. The absence of its roots and the eroded condition of its occlusal surface limit analysis to internal structures and overall shape of the tooth crown. Dental outline analysis provides a framework in which teeth can be analyzed independent of their absolute size, presence of the dental root and to a certain degree occlusal wear (see, e.g., Benazzi et al. 2011a, 2012). On other tooth types the cervical and crown outline analyses were successfully used to distinguish between Neanderthals and *Homo sapiens* as well as between fossil *Homo sapiens* and recent populations (e.g., Benazzi et al. 2011a, 2011b; Harvati et al. 2015). For the analysis of the cervical outline, the preservation of the Megalopolis molar would require reconstruction to avoid introducing a possible source of error because parts of this outline are missing. In contrast, an analysis of the crown outline is possible without reconstruction. In this chapter, we show that the method of crown outline analysis can be applied on upper third molars and thereby help shed light on the taxonomic status of the Megalopolis molar.

MATERIAL AND METHODS

Crown outline analysis on upper third molars

Tooth outlines can be analyzed by direct extraction of diameters, diagonals, and area or by geometric morphometric analysis of landmark data collected on the outline. We analyzed landmark data collected on the crown outline of the Megalopolis molar to allow the consideration of its shape as well as its form (defined as shape considered together with size). It is important to note that landmark data collected on outlines do not strictly fall into the classical concepts of fixed landmarks and semi-landmarks because of the lack of homologous fixed points and start or end points of the outline curve, respectively (for discussion of landmark types see, e.g., Bookstein 1991; Gunz et al. 2005). Therefore, the required removal of orientation, location and absolute size from the landmark coordinates cannot be achieved by a Generalized Procrustes analysis (GPA) alone. A geometric morphometric analysis of a dental outline requires additional specific steps during data collection to remove orientation and ensure homology between landmark configurations (Bauer et al. 2018; Harvati et al. 2015).

Prior to the data collection, consistent orientation was identified as possible source of error due to the high intraspecific variation of third molars (e.g., Gómez-Robles et al. 2012; Schneider et al. 2014). We used only upper third molars (M^3) with a mesial contact facet to the second molar or with a very distinct crown that immediately allowed the identification of the mesial side to minimize this source of error. Landmark data of the crown outline were collected by two observers (C.R.; J.Z.) on 39 μCT scans from original specimens with their resolution ranging from 10.3 to 36.3 μm (Table 1). The M^3 landmark data showed a very high interspecific homogeneity. Almost all individuals exhibit differences of less than 0.05 mm in their centroid size (CS) corrected landmark coordinates. In addition, one modern human specimen (Tunisia 80) was digitized five times each by two observers (C.R.; J.Z.) over the period of six months in order to evaluate intra- as well as interobserver error.

Dental casts are a source of information often neglected in geometric morphometric studies of dentition. Especially in cases when the access to CT scans is restricted, casts can provide a valuable addition to the comparative sample. Landmark data were collected by one observer (C.R.) on seven μCT scans of high resolution dental casts from fossil individuals (resolution ranging from 50.3 to 77.6 μm; Table 2). Only casts with clearly visible cervical lines were scanned and included in our analysis. To evaluate inter-method error, a high resolution dental cast of one modern human individual from the μCT scan sample (Vaihingen 13156) was created. Subsequently the cast was both μCT scanned and surface scanned. Landmark data were collected five times each from the μCT scan of the original tooth, the μCT of the cast and from the surface scan of the cast over a period of six months by one observer (C.R.).

Table 1.
Sample of μCT of original specimens in detail.

Species/Population		Individuals/ Collection numbers	Right or Left Side	scan resolution (μm)	Scanner	Collection/ Repository	Used in analysis
Homo sapiens	Neolithic from Egypt	1290	left	36.3	μCT GE Phoenix v|tome|x s240 at the University of Tübingen, Germany	Osteological Collection, University of Tübingen, Germany	all
		1299		24.5			
		1306		28.9			
	Bronze Age from Tunisia	83, 84, 85	left	23.4			
		80, 81, 82, 86	right				
	Recent from Oceania	4249, 4265, 4300	left	20.9			
		4258, 4262		23.7			
		4259	right				
		4260		22.6			
	Recent from Europe	13156, 13162, 13253, 13266	left	25.6			
		13181, 13231, 13273	right				
	fossil	La Rochette	left	10.3			
Homo neanderthalensis		Feldhofer Grotte (Fe)	left	10.3	not further specified μCT scanner	NESPOS online database	all
		Krapina (Kr) d97		16.2			
		Kr d173		32.5			
		Kr d180		20			
		Kr: d58, d163		31.2			
		Kr d109	right	18.7			
		Kr d162		27.5			
		Kr: d170, d178		25			
Homo erectus		Sangiran (Sa):NG0802.1 (Zanolli 2013, 2015)	left	20.9	μCT X8050-16 from Viscom AG at the University of Poitiers, France	Balai Pelestarian Situs Manusia Purba of Sangiran, Java	all
		Sangiran (Sa): 7-17	right	17	μCT GE Phoenix v|tome|x s240 at the University of Tübingen, Germany	Senckenberg Institute Frankfurt, Germany	
Homo heidelbergensis		Steinheim	left	25.6	μCT BIR SN001 ACTIS5 at the MPI EVA, Leipzig, Germany	Staatliches Museum für Naturkunde Stuttgart, Germany	
		Megalopolis	left	24.3	μCT GE Phoenix v|tome|x, Phoenix Service Center in Stuttgart, Germany	Museum of Palaeontology and Geology, Athens, Greece	projected into PCA

Table 2.
Sample of µCT of dental casts in detail.

Species/Population		Individuals/ Collection numbers	Right or Left Side	scan resolution (µm)	Scanner[1]	Collection/ Repository	Used in analysis
Homo sapiens	Recent from Europe	13156	left	> 50	Artec Space Spider handheld 3D surface scanner	Osteological Collection, University of Tübingen, Germany	only error calculations
				75.9	µCT GE Phoenix v|tome|x s240		
	Fossil	Brno 1 (Br); Ohalo 2 (Oh)	right		µCT GE Phoenix v|tome|x s240	dental cast collection from Dr. Sireen El Zaatari	all
		Qafzeh 9 (Qa)	left				
Homo neanderthalensis		Amud 1 (Am)	left				
		Le Petit-Puymoyen 2 (Pe)		50.3			
		Saint Césaire 1 (Sc)	right	75.9			
		Spy 1 (Sp)		77.6			

[1] All specimens were scanned at the Paleoanthropology High-Resolution Computing Tomography Laboratory, University of Tübingen, Germany.

The following protocol includes all data collection steps necessary for geometric morphometric analysis of the crown outline (Fig. 2). All teeth from the right side were mirror-imaged and treated as teeth from the left side in all subsequent steps. Mirroring of anatomical antimeres is often used to expand sample size (e.g., Bauer et al. 2018; Harvati et al. 2015). It has to be noted that combining right and left teeth might increase noise, since fluctuating asymmetry is the norm in dentition (e.g., Scott et al. 2018; Sprowls et al. 2008). An orientation system based on the cervical line ensured repeatability of the spatial positioning and orientation (Benazzi et al. 2009). A best-fit plane along the cervical line was calculated and the tooth virtually sectioned along this plane. The best-fit cervical plane was translated to the x-y plane of a coordinate system to establish a relationship between the measured crown outline and the cervical plane. In addition, this enabled a consistent orientation of the teeth. Each tooth was rotated until the mesio-distal axis was parallel to the x-

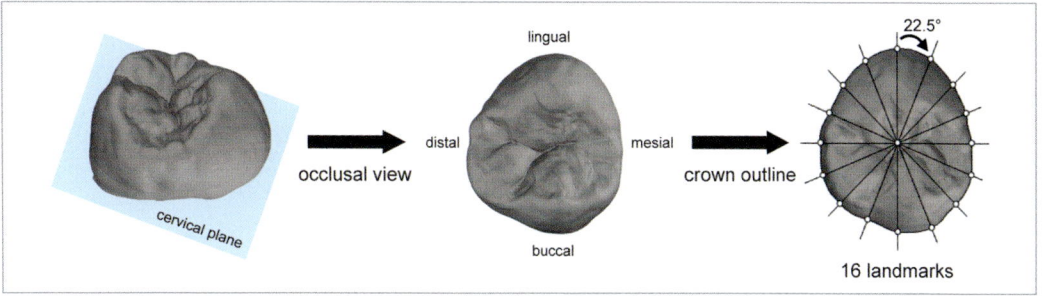

Fig. 2.
Illustration of the method used. From left to right: virtual surfaces were sectioned along a best-fit plane of the cervical line; the created cervical plane of the crown was translated to the x-y plane of a coordinate, the crown was rotated until in occlusal view the mesio-distal axis was parallel to the x-axis, the crown outline was projected onto the x-y plane and outgoing from the outline's centroid cut by 16 radii in a 22.5° angle to each other, the points of interception between the radii and the crown outline created 16 landmarks. Illustration created in Photoshop CS 5 based on virtual surface models from Rhinoceros 6.

axis with the mesial side pointing towards higher values along the x-axis. A standardized occlusal view with a 90° angle to the x-y plane was used to project the crown outline onto the x-y plane (Benazzi et al. 2011a, 2012). The projected outline's area centroid was calculated and translated to a predetermined point, here 10,10,0. Sixteen radii were digitized at an angle of 22.5° to each other outgoing from the centroid (Bauer et al. 2018; Benazzi et al. 2012). The points of interception between the radii and the crown outline created a set of 16 two-dimensional landmarks per tooth. The landmark set was statistically analyzed after scaling it to centroid size (CS) and removing location from the scaled landmark data. The CS is calculated as the square root of the summed squared of each landmark-centroid distance (Zelditch et al. 2012). A partial GPA with inhibited rotation was performed to remove scale and location at the same time. All steps of data collection were carried out in the software environments of Avizo 9.2 (FEI Visualization Sciences Group) and Rhinoceros 6 (Robert McNeel and Associates, Seattle, WA). Statistical analyses were carried out in R (R Development Core Team 2011) using published and freely available code (GPA: Morpho package; following error calculations: Morpho, geomorph, and stats packages).

The relative reproducibility of individual landmarks was assessed by calculating the error in percent of the Euclidean distance (ED) between the configuration centroid and repeat measures of each landmark in the configuration (Fig. 3a; Singleton 2002). The EDs were calculated based on the raw, not scaled landmark configurations. For each observation the configuration's centroid and the 16 ED's between the centroid and each landmark were computed. Percentage error was calculated for each landmark, and within and between observers the average deviation was determined. The measured error was below five percent in all cases, which is

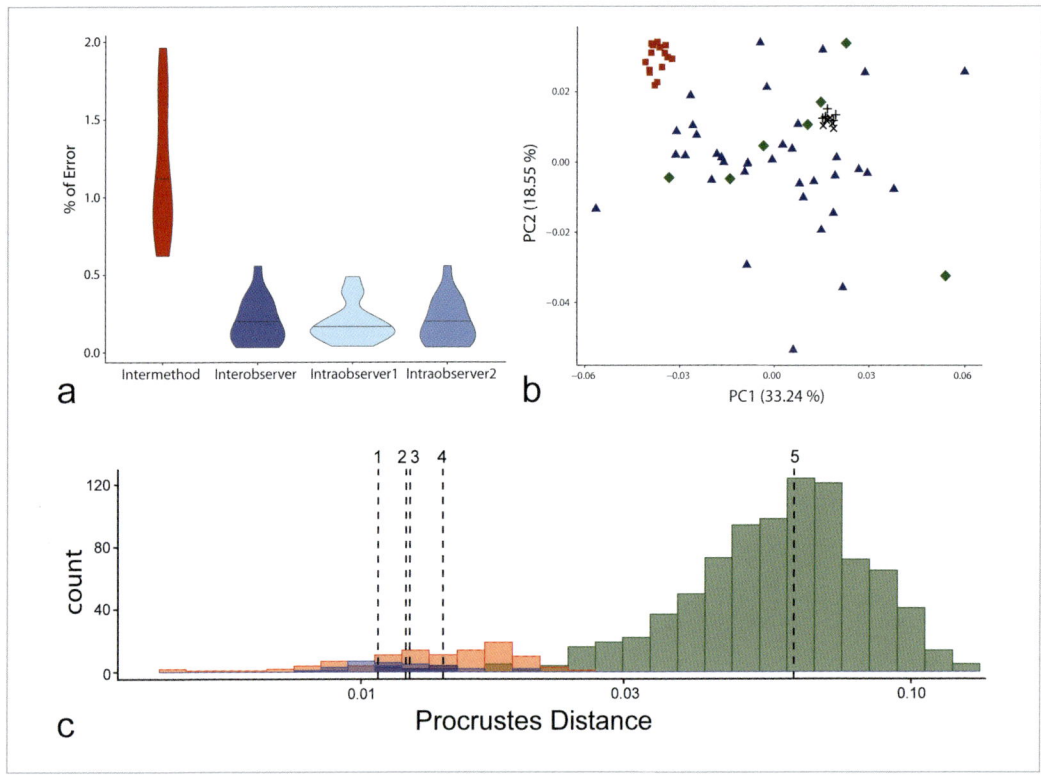

Fig. 3.
Illustrations for the error calculations. A) Error in Euclidean distance (ED) per landmark between repeated measurements. The 16 ED's are summarized for the intermethod error in red (mean = 1.173, max = 1.964), interobserver error in dark blue (mean = 0.198, max = 0.558), intraobserver error 1 in light blue (mean = 0.190, max = 0.487), and intraobserver error 2 in medium blue (mean = 0.198, max = 0.557). B) Projection of PCA of the crown outline in shape space. PCA calculated based on CT scans of original specimens; repeated measurements and data from casts later projected into the plot. CT scans shown as blue triangles, dental casts as green diamonds, intermethod error as red squares and the intraobserver error as black + and X, respectively. C) Histogram of pairwise Procrustes distances (PD). PDs along x-axis logarithmized and dotted lines showing the means. Pairwise PDs between intraobserver measurements shown in light blue (mean$_1$ = 0.011, mean$_2$ = 0.012), between the interobserver measurements in dark blue (mean$_4$ = 0.012), between the intermethod measurements in red (mean$_3$ = 0.014), and between individuals in the comparative sample shown in green (mean$_5$ = 0.061). Graphic created in R and processed in Adobe Illustrator CS5.

commonly seen as the maximum of acceptable deviation between repeated measurements. The two landmark positions with the smallest ED to the configuration's centroid showed the highest error percentages in all cases, inter-method, intraobserver as well as interobserver error. This is considered a side effect when the configuration deviates from a circular or spherical shape and of small landmark configurations (von Cramon-Taubadel et al. 2007).

The effects of inter- and intraobserver as well as inter-method error were explored in a comparative sample composed of the above mentioned sample of μCT scans (Fig. 3b-c). Pairwise Procrustes distances (PD) between multiple measurements of the same individual were compared to interindividual distances assuming that specimen affinity is not influenced when the largest PD between repeated measurements of the same individuals is smaller than the smallest PD between different specimens (Neubauer et al. 2009). The PDs were calculated based on the scaled and GPA superimposed landmark configurations. Due to the very homogenous sample, the interobserver and inter-method error overlapped with the smallest distances between individuals (Fig. 3c). The highest PDs between repeated inter-method measurements were found in surface scans. Pairwise PDs between measurements on surface scans and the comparative sample were smaller than PDs to other inter-method measurements on CT scans. In contrast, all inter- as well as intraobserver measurements showed smaller PDs to each other than to other individuals. Both intraobserver errors showed values smaller than the smallest interindividual PD. All mean errors were more than four times smaller than the interindividual distance mean.

When projecting the repeated interobserver measurements into a Principal Component Analsysis (PCA) in shape space the measurements clustered together (Fig. 3b). No other individual plotted into the space between the repeated interobserver measurements. Likewise, the inter-method measurements clustered together (Fig. 3b). The higher error in surface scans was reflected by some measurements being more scattered. A possible explanation is the less secure identification of the cervical line on the surface scans due to problems in capturing the lower part of the crown during the surface scanning. Therefore, surface scans of dental casts were excluded from further analysis due to their potential influence on the results.

Crown outline analysis of the Megalopolis molar

The cervical line on the Megalopolis molar shows damage especially on the lingual and distal sides (Fig. 1). Contrary to the illustrated surface model, the μCT scan showed the original boundary between dentin and enamel on the majority of the tooth. A best-fit cervical plane was therefore computed based on this visible part of the original and not the damaged enamel-dentin boundary.

The crown outline of the Megalopolis molar was compared to a sample of modern as well as fossil *Homo sapiens*, Neanderthals, two *Homo erectus* individuals from Sangiran and the *Homo heidelbergensis* individual from Steinheim (Tables 1, 2). The majority of the comparative sample consisted of CT scans of the original specimens with the addition of seven μCT scans of high resolution dental casts.

Multivariate Statistics

All following multivariate statistics were calculated in R (R Development Core Team 2011, packages: Morpho, geomorph, stats, MASS). The PCA is a method to reduce high-dimensional space to interpret large-scale trends of data and is subject to mathematical assumptions (Abdi and Williams 2010). For the shape PCA only the scaled and superimposed landmark coordinates were used, whereas for the form PCA, CS was added as variable. The most important assumption of this method is that the dataset does not contain outliers or influential individuals. This assumption was tested on the comparative sample (the Megalopolis molar was not used to calculate the PCAs but was projected into the plots). Cook's distance was estimated for each individual and influential individuals were identified by using the cutoff values recommended by Bollen and Jackman (1985). One individual, Qafzeh 9, reached the sensitive cutoff of $4/N$, here $N = 45$ and $\alpha = 0.05$, in shape space as well as form space. An omission did not alter the pattern of results and the individual was not excluded from the analyses so as not to further limit our already small fossil *Homo sapiens* sample. In addition, the PC scores, including those of Qafzeh 9, did not show any outliers when using the ± 3 standard deviations criterion.

Shape changes along the PCs (Figs. 4, 5) were visualized as landmark configurations at ± 2 standard deviations (sd). The landmark configurations were calculated by rotating and translating PC-scores derived from shape data back into configuration space. Therefore, the coefficients of the PC, which express the relationship between the PC and the original variables, were used to predict a hypothetical landmark configuration outgoing from a PC score ± 2 sd from the PC mean. PC-scores were converted to landmark coordinates in R (R Development Core Team 2011, packages: Morpho). Convex hulls were calculated around the extreme points of each defined group and contain no information about confidence intervals.

To further explore the relationship between shape and size in our sample, a linear regression between shape and logarithmized CS was calculated. Due to its influential Cock's distance, Qafzeh 9 was excluded for this analysis. To maximize sample size, all superimposed landmark coordinates from *Homo sapiens* and Neanderthals were pooled ($N = 41$). The regression was calculated in R with a function that performs statistical assessment based on Procrustes distances among specimens, rather than explained covariance matrices among variables (Adams et al. 2020: proc.lm function).

RESULTS

The first two shape PCs explained 55.19% of variance and their combination showed no clear separation between groups (Fig. 4), a pattern which is repeated by all higher PC's. PC1 explained 35.23% of variance

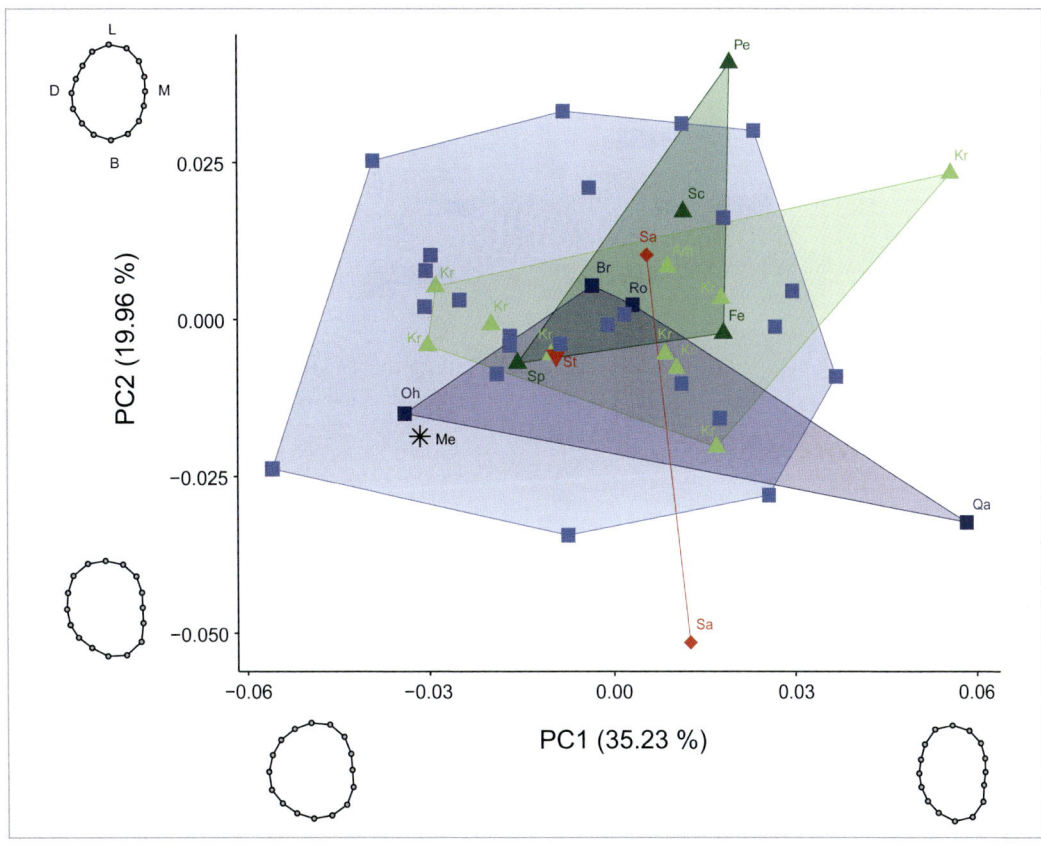

Fig. 4.
PCA of the crown outline with PC1 and PC2 projected into shape space. Modern *Homo sapiens* shown as light blue squares (N = 24), fossil *Homo sapiens* as dark blue squares (N = 4), *Homo neanderthalensis* from the Near East and the Balkan as light green triangles (N = 10), *Homo neanderthalensis* from Central Europe as dark green triangles (N = 4), *Homo erectus* as red diamonds (N = 2) and *Homo heidelbergensis* as red inverted triangle (N = 1). Megalopolis molar, shown as black star, was projected into the plot calculated based on the comparative sample. Shape changes along PCs illustrated as landmark configurations at ± 2 sd. Abbreviations of all fossil individuals listed in Tables 1, 2. Graphic created in R and processed in Adobe Illustrator CS5.

and summarized shape changes ranging from a bucco-lingual elongated oval outline with parallel mesial and distal sides (positive values) to a bucco-lingual compressed rounder outline with an outward bulging of the distal side (negative values). Teeth expressing a more positive value showed a reduction of the distal cusps on the occlusal surface compared to teeth with more negative values that showed four well developed cusps. PC2 explained 19.96% of variance and described shape changes from an outline with bulging on the lingual part of the distal side (negative values) to an outline with bulging on the buccal part of the lingual side (positive values). A reduction of the hypocone relative to the

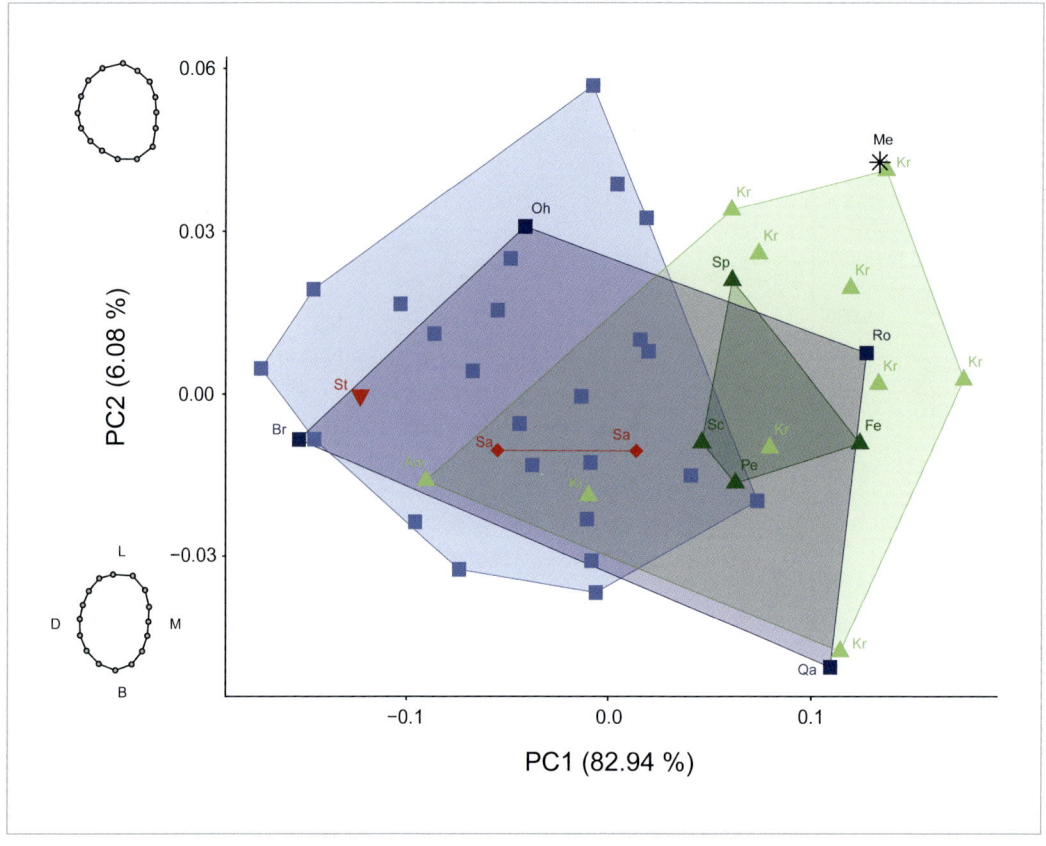

Fig. 5.
PCA of the crown outline with PC1 and PC2 projected into form space. Modern *Homo sapiens* shown as light blue squares (N = 24), fossil *Homo sapiens* as dark blue squares (N = 4), *Homo neanderthalensis* from the Near East and the Balkan as light green triangles (N = 10), *Homo neanderthalensis* from Central Europe as dark green triangles (N = 4), *Homo erectus* as red diamonds (N = 2) and *Homo heidelbergensis* as red inverted triangle (N = 1). Megalopolis molar, shown as black star, was projected into the plot calculated based on the comparative sample. PC1 negatively correlated with logarithmic centroid size. Shape changes along PC2 illustrated as landmark configurations at ± 2 sd. Abbreviations of all fossil individuals listed in Tables 1, 2. Graphic created in R and processed in Adobe Illustrator CS5.

metacone was expressed by teeth showing positive values. In contrast, a reduction of the metacone relative to the hypocone was expressed by teeth showing negative values. When projecting the Megalopolis molar into the PCA plot, it plotted among the negative values of both PC1 and PC2, reflecting its four developed cusps with a slight reduction of the metacone. The Megalopolis molar plotted into the modern *Homo sapiens* convex hull and close to the fossil *Homo sapiens* from Ohalo II.

The form PCA combines the scaled, superimposed landmark data with the variable of CS (Fig. 5). PC1 explained 82.94% of variance and was highly positively correlated with CS. PC2 explained 6.08% of variance and summarized shape changes ranging from a bucco-lingual elongated oval outline with parallel mesial and distal sides (negative values) to a bucco-lingual compressed rounder outline with an outward bulging of the distal side especially disto-lingual (positive values). Teeth expressing more negative values showed the reduction of both distal cusps while more positive values showed four well developed cusps with a slightly more pronounced hypocone relative to the metacone.

On the one hand, PC1 in form space reflected a high interspecific homogeneity by explaining 82.94% of variation. On the other hand, PC1 separated most of the Neanderthals individuals with positive values from the modern *Homo sapiens* with rather negative values. Overall, the PCA plot showed varying degrees of overlap between all groups and large intraspecific variation. Two fossil *Homo sapiens*, La Rochette and Qafzeh 9, showed the most positive values along PC1 for all *Homo sapiens* and thereby, plotted closer to Neanderthals than other *Homo sapiens*. The Neanderthal individuals Amud 1 and Krapina d097 plotted within the modern *Homo sapiens*. The chronologically older individuals from Steinheim and Sangiran plotted into the modern *Homo sapiens* variation. When projecting the Megalopolis molar into the PCA plot, it plotted in the positive values of PC1 and PC2, and thereby away from the *Homo sapiens* convex hulls. Krapina d178 plotted closest to the Megalopolis molar and showed the greatest resemblance in overall shape based on PD. Their pairwise PD was in a similar order of magnitude as the reported mean interobserver error. The Megalopolis molar would still plot outside of the *Homo sapiens* convex hulls and closest to Neanderthals from Krapina when assuming a circular error range with the plotted distance to Krapina d178 as radius.

CS in our sample ranges from 17.71 to 25.08 (Table 3), with Neanderthals on average showing higher values (mean: 22.45 ± 1.49) than *Homo sapiens* (mean: 20.63 ± 1.52). Megalopolis showed a CS of 24.08. No statistically significant relationship between shape and size was found ($F = 1.14$, $Dof = 1$, $R^2 = 0.03$, $p = 0.29$; $\alpha = 0.05$).

DISCUSSION

A high degree of variability in distal maxillary molars in both fossil as well as recent populations is repeatedly reported in the literature (e.g., Bailey 2002; Macho and Moggi-Cecchi 1992; Martinón-Torres 2006). This observation was reflected in the high intraspecific variation in shape space (Fig. 4). Nevertheless, the size corrected superimposed landmark coordinates of many individuals showed differences of less than 0.05 mm. On the one hand, it implies that the crown outline alone does not capture certain aspects of the crown morphology, e.g., the position of fissures and cusps. On the other hand, it implies that even these subtle differences can potentially be informative and therefore error measurements should not be purely based on Euclidean distances (ED). The tra-

Table 3.
Centroid sizes of all *Homo sapiens* (N = 28) and Neanderthals (N = 14).

Species	Individuals/ Collection numbers	Centroid Size[1,2]	Mean[2]	Standard deviation[2]
Homo sapiens	13273	17.71	20.63	1.52
	Brno 1	18.04		
	84	18.19		
	85	18.21		
	13181	19.02		
	13162	19.08		
	1290	19.32		
	13231	19.50		
	4300	19.66		
	83	19.94		
	81	20.11		
	4259	20.14		
	1299	20.23		
	Ohalo 2	20.25		
	4262	20.77		
	1306	20.77		
	4265	20.83		
	13266	20.84		
	13253	20.85		
	4249	20.94		
	13156	21.20		
	82	21.39		
	4260	21.48		
	86	21.50		
	80	21.88		
	4258	22.60		
	Qafzeh 9	23.39		
	La Rochettte	23.87		
Homo neanderthalensis	Amud 1	19.21	22.45	1.49
	Krapina d097	20.81		
	Saint Césaire	22.02		
	Spy 1	22.37		
	Le Petit-Puymoyen 2	22.39		
	Krapina d173	22.39		
	Krapina d162	22.68		
	Krapina d170	22.73		
	Krapina d178	23.50		
	Krapina d163	23.72		
	Feldhofer Grotte	23.80		
	Krapina d058	24.02		
	Krapina d180	24.17		
	Krapina d109	25.08		
	Megalopolis	24.08		

[1] sorted from minimum to maximum centroid size within each species.
[2] values rounded to two decimals.

ditionally accepted error range for EDs is up to five percent. In case, of M³s five percent deviation between repeated measurements could be up to four times higher than the difference between individuals at this one landmark position.

Besides high intraspecific variability, an expanded fossil sample spanning Australopithecines to recent *Homo sapiens* showed that a hypocone reduction in M³s characterizes later *Homo*, like *Homo sapiens* and Neanderthals (Gómez-Robles et al. 2012). In addition, Gómez-Robles et al. (2012) found a higher level of metacone reduction in *Homo sapiens* than in Neanderthals. We could not analyze hypocone reduction to the same level of resolution, as our sample almost exclusively consists of later *Homo*. The PCA in shape space (Fig. 4) showed varying degrees of hypocone and metacone reduction for all groups. In our sample, only six *Homo sapiens*, Sangiran 7-17, and Krapina d170, as well as Megalopolis, show signs of metacone reduction. Ohalo II, Krapina d170 and Megalopolis express almost identical values along PC2, which described the metacone reduction. It has to be noted that Gómez-Robles et al. (2012) analyzed a different landmark set which provided additional information about the occlusal surface that is absent in our analyses. Further differences in methodology include the data acquisition from photographs and the use of sliding curve semi-landmarks by Gómez-Robles et al. (2012). A curve requires a homologous point, which provides the start for a predefined number of equidistantly spaced semi-landmarks on the curve (e.g., Gunz et al. 2005). Identification of the homologous start point is highly dependent on the orientation, here the position of the tooth during photographing, as well as preservation and occlusal wear. However, the preservation of the Megalopolis molar does not allow a secure identification of the landmarks on the occlusal surface nor a homologous start point for a curve of semi-landmarks.

Beyond shape differences, dental size, and especially linear measurements are commonly used to discriminate between both fossil and recent populations (e.g., Harvati et al. 2003, 2013; Smith et al. 2015; Xing et al. 2014). Xirotiris et al. (1979) reported a mesio-distal breadth 9.1 mm and a bucco-lingual length of 10.3 mm for the Megalopolis molar, which was at the lower end of their measurements for Neanderthals and *Homo sapiens*. In contrast, when using CS as a measure of size, the Megalopolis molar falls slightly outside the range of variation of our *Homo sapiens* comparative sample and at the upper end of the Neanderthal range (Table 3). The difference is that CS is based on multiple aspects (landmark coordinates) of the tooth while a linear measurement captures a single aspect of the tooth. For example, mesio-distal and bucco-lingual measurements hardly take into account reductions of one distal cusp. Nevertheless, both linear measurements and CS show the same trend that, on average, Neanderthals have larger M³s than *Homo sapiens* (e.g., Harvati et al. 2013; Xirotiris et al. 1979).

Macho and Moggi-Cecchi (1992) suggested that simplification in M³ morphology is partially caused by a reduction in size. The PCA in form space (Fig. 5) showed no relationship between shape and size in our sample of M³s. Larger teeth (PC1 positive values) varied in their morphology from completely reduced distal cusps to four well-developed cusps with only a slight reduction of the metacone. This reflects a common issue regarding allometric relationships in the hominin dentition. On the one hand, some studies have assumed or established no clear allometric effects in hominin dentition (e.g., Wood et al. 1983; Bailey and Lynch 2005). On the other hand, additional studies have suggested small but significant allometric relationships (e.g., Martinón-Torres et al. 2006; Gómez- Robles et al. 2008). It can be assumed that allometric effects do not explain morphological changes over an evolutionary time span, but might have some impact on patterns of intraspecific morphological variation (Gómez-Robles et al. 2012).

The process of dental development is multifactorial and thereby dental morphology is influenced by genetic, epigenetic and environmental factors (Brook et al. 2014a, 2014b). Genetic admixture might partially explain the observed overlap between *Homo sapiens* and Neanderthals in form space (Fig. 5). The two fossil *Homo sapiens* La Rochette and Qafzeh 9 plotted closer to Neanderthals than other *Homo sapiens* while the Neanderthal individuals Amud 1 and Krapina d097 plotted within the modern *Homo sapiens*. Genetic evidence suggests admixture between Neanderthals and *Homo sapiens* in the Middle and Late Pleistocene (e.g., Green et al. 2010; Posth et al. 2017; Sankararaman et al. 2012). Therefore, admixture cannot be ruled out entirely as a possible explanation for the observed morphology of La Rochette, Qafzeh 9, Krapina d097 and Amud 1. Admixture and genetics are based on the heritability of traits. Biological distance studies commonly use dental traits and often assume equal and additive inheritance of traits (e.g., Macchiarelli et al. 2008; Vargiu et al. 2009). However, non-additive genetic variation might preserve certain dental traits over time (Edgar and Ousley 2016). The chronologically older individuals from Steinheim and Sangiran plotted into the modern *Homo sapiens* variation or express an even more extreme metacone reduction (Figs. 4, 5). On the one hand, this might imply that some aspects of the crown outline in *Homo sapiens* are conserved and resemble the primitive state found in chronologically older groups. On the other hand, this might be an artifact of the high intraspecific variation in combination with the underrepresentation of *Homo heidelbergensis* and *Homo erectus* in our sample.

All in all, the two main components of the crown outline shape of the Megalopolis molar matched the variation found in the Holocene comparative sample. In contrast, outline form and overall shape did not match the Holocene sample. The form PCA, the Procrustes distances based on overall shape, as well as its centroid size, grouped the Megalopolis molar with the Neanderthal comparative sample. Although our samples were small, we cautiously interpret these results as indicating that the Megalopolis specimen likely dates to the Pleistocene and has affinities with the Neanderthal lineage.

It is important to note that our comparative sample lacked important individuals from the Middle Pleistocene of Europe (e.g., Petralona and Sima de los Huesos) and Africa (e.g., Broken Hill and Herto) and also many Neanderthal specimens. These were not possible to include due to a lack of access to dental casts or CT scans. A secure classification of the Megalopolis molar would require a more comprehensive sampling framework for the taxonomic interpretation and must be tested with further analyses and with an expanded comparative fossil sample.

The number of Neanderthal and pre-Neanderthal fossils from Greece has remained small even while the number of archaeological sites from the Middle and Lower Paleolithic has increased over recent years (e.g., Harvati 2016; Tourloukis and Harvati 2018). The oldest, radiometrically dated, Pleistocene site in Greece known to date, Marathousa 1, is located within the Megalopolis Basin. Marathousa 1 was dated to 400-500 ka (Blackwell et al. 2018; Jacobs et al. 2018) and provides a rich lithic as well as faunal assemblage (e.g., Tourloukis et al. 2018a; Konidaris et al. 2018). In contrast, the oldest human fossil, Petralona, commonly attributed to *Homo heidelbergensis*, or pre-Neanderthal, is not well dated (Dean et al. 1998; Hublin 1998, 2009). A Neanderthal presence in Greece was demonstrated by several Middle Paleolithic find spots and sites (Tourloukis and Harvati 2018), especially in the Mani Peninsula, Southern Peloponnese, where human fossil have also been recovered (e.g., Elefanti et al. 2008; Tourloukis et al. 2016). Three sites in Mani yielded Neanderthal remains: Lakonis (Harvati et al. 2003), Kalamakia (Harvati et al. 2013) and Apidima (Pitsios 1999; Harvati et al. 2019), dated to ca. 40 ka, between 100 and 40 ka, and to ca. 170 ka, respectively.

Although a direct or indirect dating is not available for the Megalopolis molar, the geology of the Megalopolis Basin suggests a Middle Pleistocene geological age, under the assumption that the molar is a fossil and not a modern intrusion. The Megalopolis Basin is a tectonic half-graben and filled with Neogene to Holocene sediments (Vinken 1965). Geological mapping showed that the surrounding hills consist of pre-Pliocene basement while the majority of the basin encompasses sediments of Pleistocene, especially Middle Pleistocene, origin (Siavalas et al. 2009). The Middle Pleistocene sediments can be divided into the Megalopolis member of the Choremi formation, which consists of fluvial deposits, and the Marathousa member, which consists of fossil-rich lacustrine deposits (Löhnert and Nowak 1965; Vinken 1965). On the grounds of paleomagnetic, cyclostratigraphic, biochronological and palynological data, the lacustrine sequence has been chronologically bracketed between ca. 950-350 ka (van Vugt et al. 2000; Okuda et al. 2001), or ca. 800-300 ka (Tourloukis et al. 2018b), with its upper age-limit being poorly constrained at around 300 or 200 ka (see also Jacobs et al. 2018; Blackwell et al. 2018). Recent multiproxy paleoenvironmental reconstruction (Bludau et al. 2021) has highlighted the role of the Megalopolis Basin as a potential glacial refugium for Pleistocene humans due to its ability to retain freshwater bodies during glacial periods. The Megalopolis molar was part of a surface collection, in which many of the collected fossils were found still embedded in blocks of lacustrine sediments and derived from deposits of the Marathousa Member (Sickenberg 1976: 26). Therefore, we hypothesize that the Megalopolis molar dates to the Middle Pleistocene and derives from the Marathousa Member of the Choremi Formation.

Future work should focus on expanding the comparative sample, especially the fossil sample of Neanderthals in order to span their entire spatial and temporal range, as well as representatives of *Homo heidelbergensis*, which are underrepresented in our study. Methods based on segmentation, e.g., analyses of the enamel dentine junction (EDJ), are limited by the state of preservation of the Megalopolis molar, which complicate the differentiation between enamel and dentine. Future improvements in CT scanning techniques might enable the latter and provide a more complete picture of the taxonomic affinities of the Megalopolis molar.

CONCLUSION

The case study of the Megalopolis molar demonstrates the necessity of analytical tools that allow the study of incomplete specimens. The method of crown outline analysis was applied to M^3s and allowed the first quantitative study of the Megalopolis molar. On the basis of our results we conclude that the Megalopolis molar most likely represents a Pleistocene specimen with Neanderthal lineage affinities. The importance of the Megalopolis molar is threefold. First, it contributes to the Pleistocene human fossil record of Greece. Every new individual is valuable in adding to our understanding of human evolution in this relatively understudied region (see, e.g., Harvati 2016; Tourloukis and Harvati 2018). Second, the Megalopolis molar was found in the Megalopolis Basin, where the Middle Pleistocene site Marathousa 1 is also located. Both this specimen and the site highlight the potential of this region for yielding precious paleoanthropological finds. Furthermore, the present study adds to the examples of methodological improvements enabling new insights from known material, which could not be studied at the time of discovery due to its fragmentary status or taphonomic distortion.

ACKNOWLEDGMENTS

We would like to thank Catherine C. Bauer and Vangelis Tourloukis for their useful comments and remarks about the methodology and geology of the Megalopolis Basin, respectively. We are grateful to Sireen El Zaatari for access to her collection of high resolution dental casts and her valuable comments, and to Panagiotis Karkanas for assistance with photographing the specimen. For access to the Megalopolis molar we thank the Faculty of Geology and Geoenvironment and the Museum of Paleontology and Geology, National and Kapodistrian University of

Athens, as well as Vassilis Karakitsios, George Lyras, Hara Drinia and the late Nikolaos Symeonidis. We would like to acknowledge all institutions and researchers that provided CT scans for this research: NESPOS online database; the Staatliches Museum für Naturkunde Stuttgart for access to the Steinheim material; the Senckenberg Research Institute in Frankfurt am Main and Friedemann Schrenk for access to Sangiran 7-17; the Balai Pelestarian Situs Manusia Purba of Sangiran, Java, as well as Arnaud Mazurier, Clément Zanolli, Roberto Macchiarelli, Harry Widianto, Dominique Grimaud-Hervé and Françoise Sémah for access to Sangiran NG0802.1. In addition, we would like to acknowledge the assistance of the Paleoanthropology High Resolution Computing Tomography Laboratory at the Eberhard-Karls University of Tübingen, supported by the DFG INST 37/706-1, in scanning the dental casts and all used individuals from the osteological collection of the University of Tübingen. This research was supported by the European Research Council (ERC CoG no. 724703). We are grateful to the reviewers, Mirjana Roksandic and Hugo Reyes-Centeno, for their comments and suggestions.

REFERENCES

Abdi, H., and L. Williams. 2010. Principal component analysis. *Willey Interdisciplinary Reviews: Computational Statistics* 2: 433-459. doi:10.1002/wics.101.

Adams, D., M. Collyer, and A. Kaliontzopoulou. 2020. Geomorph: Software for geometric morphometric analyses. Version 3.2.1. https://cran.r-project.org/package=geomorph.

Athanassiou, A. 2018. Pleistocene vertebrates from the Kyparíssia lignite mine, Megalopolis Basin, S. Greece: Rodentia, Carnivora, Proboscidea, Perissodactyla, Ruminantia. *Quaternary International* 497: 198-221. doi:10.1016/j.quaint.2018.06.042.

Athanassiou, A., D. Michailidis, E. Vlachos, V. Tourloukis, N. Thompson, and K. Harvati. 2018. Pleistocene vertebrates from the Kyparíssia lignite mine, Megalopolis Basin, S. Greece: Testudines, Aves, Suiformes. *Quaternary International* 497: 178–197. doi:10.1016/j.quaint.2018.06.030.

Bailey, S. E. 2002. Neandertal dental morphology: Implications for modern human origins. Ph.D. Dissertation, Arizona State University.

Bailey, S. E., and J. M. Lynch. 2005. Diagnostic differences in mandibular P4 shape between Neandertals and anatomically modern humans. *American Journal of Physical Anthropology* 126: 268e277.

Bauer, C. C., S. Benazzi, A. Darlas, and K. Harvati. 2018. Geometric morphometric analysis and internal structure measurements of the Neanderthal lower fourth Premolars from Kalamakia, Greece. *Quaternary International* 497: 14–21. doi:10.1016/j.quaint.2018.01.035.

Benazzi, S., M. Fantini, F. De Crescenzio, F. Persiani, and G. Gruppioni. 2009. Improving the spatial orientation of human teeth using a virtual 3D approach. *Journal of Human Evolution* 56: 286–293. doi:10.1016/j.jhevol.2008.07.006.

Benazzi, S., M. Coquerelle, L. Fiorenza, F. Bookstein, S. Katina, and O. Kullmer. 2011a. Comparison of dental measurement systems for taxonomic assignment of first molars. *American Journal of Physical Anthropology* 144: 342–54. doi:10.1002/ajpa.21409.

Benazzi, S., K. Douka, C. Fornai, C. C. Bauer, O. Kullmer, J. Svoboda, I. Pap, F. Mallegni, P. Bayle, and M. Coquerelle. 2011b. Early dispersal of modern humans in Europe and implications for Neanderthal behavior. *Nature* 479: 525–529. doi:10.1038/nature10617.

Benazzi, S., C. Fornai, L. Buti, M. Toussaint, F. Mallegni, S. Ricci, G. Gruppioni, G. W. Weber, G. W., Condemi, and A. Ronchitelli. 2012. Cervical and crown outline analysis of worn Neanderthal and modern human lower second deciduous molars. *American Journal of Physical Anthropology* 149: 537–546. doi:10.1002/ajpa.22155.

Blackwell, B. A. B., N. Sakhrani, I. K. Singh, K. K. Gopalkrishna, V. Tourloukis, E. Panagopoulou, P. Karkanas, J. I. B. Blickstein, A. R. Skinner, J. A. Florentin, and K. Harvati. 2018. ESR Dating Ungulate Teeth and Molluscs from the Paleolithic Site Marathousa 1, Megalopolis Basin, Greece. *Quaternary* 1: 22. doi:10.3390/quat1030022.

Bludau, I. J. E., P. Papadopoulou, G. Iliopoulos, M. Weiss, E. Schnabel, N. Thompson, V. Tourloukis, C. Zachow, S. Kyrikou, G. E. Konidaris, P. Karkanas, E. Panagopoulou, K. Harvati, and A. Junginger. 2021. Lake-level changes and their paleo-climatic implications at the MIS12 Lower Paleolithic (Middle Pleistocene) site Marathousa 1, Greece. *Frontiers in Earth Science* 9: 668445. doi: 10.3389/feart.2021.668445.

Bollen, K. A., and R. Jackman. 1985. Regression diagnostics: An expository treatment of outliers and influential cases. *Sociological Methods and Research* 13: 510–542. doi:10.1177/0049124185013004004.

Bookstein, F. L. 1991. *Morphometric Tools for Landmark Data: Geometry and Biology*. New York: Cambridge University Press, Cambridge (UK).

Brook, A. H., M. B. O'Donnell, A. Hone, E. Hart, T. E. Hughes, R. N. Smith, and G. C. Townsend. 2014a. General and craniofacial development are complex adaptive processes influenced by diversity. *Australian Dental Journal* 59: 13–22. doi:10.1111/adj.12158.

Brook, A. H., J. Jernvall, R. N. Smith, T. R. Hughes, and G. C. Townsend. 2014b. The dentition: The outcomes of morphogenesis leading to variations of tooth number, size and shape. *Australian Dental Journal* 59: 131–142. doi:10.1111/adj.12160.

Carter, M., and J. Shieh. 2015. *Guide to Research Techniques in Neuroscience*, pp. 117–144. Amsterdam, Boston: Elsevier Academic Press. doi:10.1016/B978-0-12-800511-8.00005-8.

Dean, D., J. J. Hublin, R. Holloway, and R. Ziegler. 1998. On the phylogenetic position of the pre-Neandertal specimen from Reilingen, Germany. *Journal of Human Evolution* 34: 485–508. doi:10.1006/jhev.1998.0214.

Duval, M., C. Falguères, and J. J. Bahain. 2012. Age of the oldest hominin settlements in Spain: Contribution of the combined U-series/ESR dating method applied to fossil teeth. *Quaternary Geochronology* 10: 412–417. doi:10.1016/j.quageo.2012.02.025.

Edgar, H. J., and S. D. Ousley. 2016. Dominance in dental morphological traits: Implications for biological distance studies. In *Biological Distance Analysis: Forensic and Bioarchaeological Perspectives*, ed. by M. A. Pilloud and J. T. Hefner, pp. 317–332. Amsterdam, Boston: Elsevier Academic Press. doi:10.1016/B978-0-12-801966-5.00017-2.

Elefanti, P., E. Panagopoulou, and P. Karkanas. 2008. The transition from the Middle to the Upper Paleolithic in the Southern Balakans: The evidence from the Lakonis 1 Cave, Greece. *Eurasian Prehistory* 5: 85–95.

Giusti, D., V. Tourloukis, G. E. Konidaris, N. Thompson, P. Karkanas, E. Panagopoulou, and K. Harvati. 2018. Beyond maps: Patterns of formation processes at the Middle Pleistocene open-air site of Marathousa 1, Megalopolis Basin, Greece. *Quaternary International* 497: 137–153. doi:10.1016/j.quaint.2018.01.041.

Gómez-Robles, A., M. Martinón-Torres, J. M. B. de Castro, L Prado, S. Sarmiento, and J. L. Arsuaga. 2008. Geometric morphometric analysis of the crown morphology of the lower first premolar of hominins, with special attention to Pleistocene Homo. *Journal of Human Evolution* 55: 627e638.

Gómez-Robles, A., J. M. B. de Castro, M. Martinón-Torres, L. Prado-Simón, and J. L. Arsuaga. 2012. A geometric morphometric analysis of hominin upper second and third molars, with particular emphasis on European Pleistocene populations. *Journal of Human Evolution* 63: 512-526. doi:10.1016/j.jhevol.2012.06.002.

Gunz, P., P. Mitteroecker, F. L. Bookstein. 2005. Semilandmarks in three dimensions. In *Modern Morphometrics in Physical Anthropology*, ed. by D. E. Slice, pp. 73–98. Boston: Springer. doi:10.1007/0-387-27614-9_3.

Green, R. E., J. Krause, A. W. Briggs, T. Maricic, U. Stenzel, M. Kircher, N. Patterson, H. Li, W. Zhai, M. H.-Y. Fritz, et al. 2010. A draft sequence of the Neandertal genome. *Science* 328:710–722. doi:10.1126/science.1188021.

Harvati, K., E. Panagopoulou, and P. Karkanas. 2003. First Neanderthal remains from Greece: The evidence from Lakonis. *Journal of Human Evolution* 45: 465–473. doi:10.1016/j.jhevol.2003.09.005.

Harvati, K., A. Darlas, S. E. Bailey, T. R. Rein, S. El Zaatari, L. Fiorenza, L., O. Kullmer, and E. Psathi. 2013. New Neanderthal remains from Mani Peninsula, Southern Greece: The Kalamakia Middle Paleolithic cave site. *Journal of Human Evolution* 64: 486–499. doi:10.1016/j.jhevol.2013.02.002.

Harvati, K., C. C. Bauer, F. E. Grine, S. Benazzi, R. R. Ackermann, K. L. van Niekerk, and C. S. Henshilwood. 2015. A human deciduous molar from the Middle Stone Age (Howiesons Poort) of Klipdrift Shelter, South Africa. *Journal of Human Evolution* 82: 190–196. doi:10.1016/j.jhevol.2015.03.001.

Harvati, K. 2016. Paleoanthropology in Greece: Recent findings and interpretations. In *Paleoanthropology of the Balkans and Anatolia: Human Evolution and Its Context*, ed. by K. Harvati and M. Roksandic, pp. 3–14. Dordrecht: Springer. doi:10.1007/978-94-024-0874-4_1.

Harvati, K., C. Röding, A. M. Bosman, A. F. Karakostis, R. Grün, C. Stringer, P. Karkanas, N. C. Thompson, V. Koutoulidis, L. A. Moulopoulos, V. G. Gorgoulis, and M. Kouloukoussa. 2019. Apidima Cave fossil provides earliest evidence of Homo sapiens in Eurasia. *Nature*. doi:10.1038/s41586-019-1376-z.

Hublin, J. J..1998. Climatic changes, paleogeography, and the evolution of the Neandertals. In *Neandertals and Modern Humans in Western Asia*, ed. by T. Akazawa, K. Aoki and O. Bar-Yosef, pp. 295–310. New York: Plenum Press. doi:10.1007/0-306-47153-1_18.

Hublin, J. J.. 2009. The origin of Neandertals. *Proceedings of the National Academy of Sciences* 106: 16022–16027. doi:10.1073/pnas.0904119106.

Konidaris, G. E., A. Athanassiou, V. Tourloukis, N. Thompson, D. Giusti, E. Panagopoulou, and K. Harvati. 2018. The skeleton of a straight-tusked elephant (Palaeoloxodon antiquus) and other large mammals from the Middle Pleistocene butchering locality Marathousa 1 (Megalopolis Basin, Greece): Preliminary results. *Quaternary International* 497: 65–84. doi:10.1016/j.quaint.2017.12.001.

Konidaris, G. E., V. Tourloukis, A. Athanassiou, D. Giusti, N. Thompson, E. Panagopoulou, P. Karkanas, and K. Harvati. 2019. Marathousa 2: A new Middle Pleistocene locality in Megalopolis Basin (Greece) with evidence of human modifications on faunal remains. *PESHE* 8: 82.

Jacobs, Z., B. Li, P. Karkanas, V. Tourloukis, N. Thompson, E. Panagopoulou, and K. Harvati. 2018. Optical dating of K-feldspar grains from Middle Pleistocene lacustrine sediment at Marathousa 1 (Greece). *Quaternary International* 497: 170–177. doi:10.1016/j.quaint.2018.06.029.

Koenigswald, W. V., and W. D. Heinrich. 2007. Biostratigraphische Begriffe aus der Säugetierpaläontologie für das Pliozän und Pleistozän Deutschlands. *Eiszeitalter und Gegenwart Quaternary Science Journal* 56: 96–115.

Landis, J., and G. Koch. 1977. The measurement of observer agreement for categorical data. *Biometrics* 33:159–174. doi:10.2307/2529310.

Löhnert, E., and H. Nowack. 1965. Die Braunkohlenlagerstätte von Khoremi im Becken von Megalopolis/Peloponnes. *Geologisches Jahrbuch* 82: 847–868.

Macchiarelli, R., L. Bondioli, and A. Mazurier. 2008. Virtual dentitions: Touching the hidden evidence. In *Technique and Application in Dental Anthropology*, ed. by J. D. Irish and G. C. Nelson, Cambridge Studies in Biological and Evolutionary Anthropology 53, pp. 426–448. Cambridge: Cambridge University Press. doi:10.1017/CBO9780511542442.018.

Macho, G. A., and J. Moggi-Cecchi. 1992. Reduction of maxillary molars in *Homo sapiens sapiens*: a different perspective. *American Journal of Physical Anthropology* 87: 151e159.

Marinos, G. 1975. Über einen menschlichen Zahn unter den Säugetier-Resten biharischen Alters von Megalopolis. *Annales Géologiques des Pays Helléniques* 27: 64–65.

Martinón-Torres, M., M. Bastir, J. M. B. de Castro, A. Gómez, S. Sarmiento, A. Muela, J. L. Arsuaga. 2006. Hominin lower second premolar morphology: Evolutionary inferences through geometric morphometric analysis. *Journal of Human Evolution* 50: 523e533.

Melentis, J. K. 1961. Die Dentition der Pleistozänen Proboscidier des Beckens von Megalopolis im Peloponnes (Griechenland). *Annales Géologiques des Pays Helléniques* 12: 153–262.

Neubauer, S., P. Gunz, and J. J. Hublin. 2009. The pattern of endocranial ontogenetic shape changes in Humans. *Journal of Anatomy* 215: 240–255. doi:10.1111/j.1469-7580.2009.01106.x.

Okuda, M., N. van Vugt, T. Nakagawa, M. Ikeya, A. Hayashida, Y. Yasuda, and T. Setoguchi. 2002. Palynological evidence for the astronomical origin of lignite–detritus sequence in the Middle Pleistocene Marathousa member, Megalopolis, SW Greece. *Earth and Planetary Science Letters* 201: 143–157. doi:10.1016/S0012-821X(02)00706-9.

Papagopoulou, E., V. Tourloukis, N. Thompson, A. Athanassiou, G. Tsartsidou, G. E. Konidaris, D. Giusti, P. Karkanas, and K. Harvati. 2015. Marathousa 1: A new Middle Pleistocene archaeological site from Greece. *Antiquity* 89: Project Gallery. doi:10.15496/publikation-5878.

Pitsios, T. K. 1999. Paleoanthropological research at the cave site of Apidima and the surrounding region (South Peloponnese, Greece). *Anthropologischer Anzeiger* 57: 1–11.

Posth, C., C. Weißling, K. Kitagawa, L. Pagani, L. Van Holstein, F. Racimo, N. J. Conard, C. J. Kind, H. Bocherens, and J. Krause. 2017. Deeply divergent archaic mitochondrial genome provides lower time boundary for African gene flow into Neanderthals. *Nature Communications* 8: 16046. doi:10.1038/ncomms16046.

R Development Core Team. 2011. *R: A language and Environment for Statistical Computing*. R Foundation for Statistical Computing, Vienna, Austria. ISBN 3-900051-07-0.

Sankararaman, S., N. Patterson, H. Li, S. Pääbo, D. Reich. 2012. The date of interbreeding between Neanderthals and modern humans. *PLOS Genetics* 8: e1002947. doi:10.1371/journal.pgen.1002947.

Schneider, T., K. Filo, A. L. Kruse, M. Locher, K. W. Grätz, and H. T. Lübbers. 2014. Variations in the anatomical positioning of impacted mandibular wisdom teeth and their practical implications. *Swiss Dental Journal* 124: 520-538.

Scott, G., C. Turner II, G. Townsend, G., and M. Martinón-Torres. 2018. *The Anthropology of Modern Human Teeth: Dental Morphology and its Variation in Recent and Fossil Homo sapiens*. Cambridge Studies in Biological and Evolutionary Anthropology 79. Cambridge: Cambridge University Press. doi:10.1017/9781316795859.

Singleton, M. 2002. Patterns of cranial shape variation in the Papionini (Primates: Cercopithecinae). *Journal of Human Evolution* 42:547–578. doi:10.1006/jhev.2001.0539.

Sickenberg, O. 1976. Eine Säugetierfauna des tieferen Bihariums aus dem Becken von Megalopolis (Peloponnes, Griechenland). *Annales Géologiques des Pays Helléniques* 27: 25–63.

Siavalas, G., M. Linou, A. Chatziapostolou, S. Kalaitzidis, H. Papaefthymiou, and K. Christanis. 2009. Palaeoenvironment of Seam I in the marathousa lignite mine, Megalopolis Basin (Southern Greece). *International Journal of Coal Geology* 78: 233–248. doi:10.1016/j.coal.2009.03.003.

Skouphos, T. G. 1905. Über die palaeontologischen Ausgrabungen in Griechenland in Beziehung auf das Vorhandensein des Menschen. In *Comptes Rendus du Congrès International d'Archéologie*, pp. 231–236. Athènes.

Smith, P., J. S. Brink, J. W. Hoffman, L. C. Bam, R. Nshimirimana, and F.C. de Beer. 2015. The late Middle Pleistocene upper third molar from Florisbad: Metrics and morphology. *Transactions of the Royal Society of South Africa* 70: 233–244. doi:10.1080/0035919X.2015.1065930.

Sprowls, M. W., R. E. Ward, P. L. Jamison, and J. K. Hartsfield. 2008. Dental arch asymmetry, fluctuating dental asymmetry, and dental crowding: A comparison of tooth position and tooth size between antimeres. *Seminars in Orthodontics* 14: 157–165. doi:10.1053/j.sodo.2008.02.006.

Thompson, N., V. Tourloukis, E. Panagopoulou, and K. Harvati. 2018. In search of Pleistocene remains at the gates of Europe: Directed surface survey of the Megalopolis Basin (Greece). *Quaternary international* 497: 22–32. doi:10.1016/j.quaint.2018.03.036.

Tourloukis, V., N. Thompson, C. Garefalakis, P. Karkanas, G. E. Konidaris, E. Panagopoulou, and K. Harvati. 2016. New Middle Palaeolithic sites from the Mani Peninsula, Southern Greece. *Journal of Field Archaeology* 41: 68–83. doi:10.1080/00934690.2015.1125223.

Tourloukis, V., N. Thompson, E. Panagopoulou, D. Giusti, G. E. Konidaris, P. Karkanas, and K. Harvati. 2018a. Lithic artifacts and bone tools from the Lower Palaeolithic Site Marathousa 1, Megalopolis, Greece: Preliminary results. *Quaternary International* 497: 47–64. doi:org/10.1016/j.quaint.2018.05.043.

Tourloukis, V., G. Muttoni, P. Karkanas, E. Monesi, G. Scardia, E. Panagopoulou, and K. Harvati. 2018b. Magnetostratigraphic and chronostratigraphic constraints on the Marathousa 1 Lower Palaeolithic site and the Middle Pleistocene deposits of the Megalopolis Basin, Greece. *Quaternary International* 497: 154–169. doi:10.1016/j.quaint.2018.03.043.

Tourloukis, V., and K. Harvati. 2018. The Palaeolithic record of Greece: A synthesis of the evidence and a research agenda for the future. *Quaternary International* 466: 48–65. doi:10.1016/j.quaint.2017.04.020.

White, S., J. A. J. Gowlett, and M. Grove. 2014. The place of the Neanderthals in hominin phylogeny. *Journal of Anthropological Archaeology* 35: 32–50.

Wood, B., S. A. Abbott, and S. H. Graham. 1983. Analysis of the dental morphology of Plio-Pleistocene hominids II: Mandibular molars - study of cusp areas, fissure pattern and cross sectional shape of the crown. *Journal of Anatomy* 137: 287e314.

Van Vugt, N., H. de Bruijn, T. van Kolfschoten, and C. G. Langereis. 2000. Magneto- and cyclostratigraphy and mammal-fauna's of the Pleistocene lacustrine Megalopolis Basin, Peloponnesos, Greece. *Geological Ultrajectina* 189: 69–92.

Vargiu, R., A. Cucina, and A. Coppa. 2009. Italian populations during the Copper Age: Assessment of biological affinities through morphological dental traits. *Human Biology* 81: 479–494. doi:10.3378/027.081.0406.

Venables, W. N., and B. D. Ripley. 2002. *Modern Applied Statistics with S.* 4th edition. New York: Springer. doi:10.1007/978-0-387-21706-2.

Vinken, R. 1965. Stratigraphie und Tektonik des Beckens von Megalopolis (Peloponnes, Griechenland). *Geologisches Jahrbuch* 83: 97-148.

von Cramon-Taubadel, N., B. C. Franzier, and M. Marizón Lahr. 2007. The problem of assessing landmark error in geometric morphometrics: Theory, methods, and modifications. *American Journal of Physical Anthropology* 134: 24–35. doi:10.1002/ajpa.20616.

Wandsnider, L. 2004. Solving the puzzle of the archaeological labyrinth: Time perspectivism in Mediterranean surface archaeology. In *Side-by-Side Survey: Comparative Regional Studies in the Mediterranean World*, ed. by S. Alcock and J. F. Cherry, pp. 49–62. Anthropology Faculty Publications 75. Oxford: Oxbow Press.

Xing, S., M. Martinón-Torres, J. M. B. de Castro, Y. Zhang, X. Fan, L. Zheng, W. Huang, and W. Liu. 2014. Middle Pleistocene hominin teeth from Longtan Cave, Hexian, China. *PloS one* 9: e114265. doi:10.1371/journal.pone.0114265.

Xirotiris, N., W. Henke, and N. Symeonidis. 1979. Der M^3 von Megalopolis – ein Beitrag zu seiner Morphologischen Kennzeichnung. *Zeitschrift für Morphologie und Anthropologie* 70: 117–122.

Zanolli, C. 2013. Additional evidence for morpho-dimensional tooth crown variation in a new Indonesian *H. erectus* sample from the Sangiran Dome (Central Java). *PLoS ONE* 8: e67233. doi:10.1371/journal.pone.0067233.

Zanolli, C. 2015. Molar Crown Inner Structural Organization in Javanese *Homo erectus*. *American Journal of Physical Anthropology* 156: 148–157. doi:10.1002/ajpa.22611.

Zelditch, M., D. Swiderski, and H. Sheets. 2012. *Geometric Morphometrics for Biologists: A Primer.* 2nd Edition. San Diego: Elsevier Academic Press.

Chapter 2

Direct U-series dating of the Apidima C human remains

Katerina Harvati[1,2,3,4], Rainer Grün[5,6], Mathieu Duval[5,7], Jian-xin Zhao[8], Alexandros Karakostis[2,4], Vangelis Tourloukis[1,4], Vassilis Gorgoulis[9,10,11], Mirsini Kouloukoussa[4,9]

Abstract

The site of Apidima, in southern Greece, is one of the most important Paleolithic sites in Greece and southeast Europe. One of the caves belonging to this cave complex, Cave A, has yielded human fossil crania Apidima 1 and 2, showing the presence of an early *Homo sapiens* population followed by a Neanderthal one in the Middle Pleistocene. Less known are the human remains reportedly recovered from Cave C at Apidima. These include a number of isolated elements, but also a partial skeleton interpreted as a female burial, Apidima 3, proposed by Pitsios (e.g., Pitsios 1999) to be associated with Aurignacian lithics and to date to ca. 30 ka. In light of the rarity of the Upper Paleolithic in Greece, and the general scarcity of human remains associated with the Aurignacian, the remains from

---

[1] Paleoanthropology, Senckenberg Centre for Human Evolution and Palaeoenvironments, Eberhard Karls University of Tübingen, Germany.
[2] DFG Center of Advanced Studies 'Words, Bones, Genes, Tools', Eberhard Karls University of Tübingen, Germany.
[3] Centre for Early Sapiens Behaviour (SapienCE), Department of Archaeology, History, Cultural Studies and Religion, University of Bergen, Norway.
[4] Museum of Anthropology, School of Medicine, National and Kapodistrian University of Athens, Greece.
[5] Australian Research Centre for Human Evolution, Griffith University, Nathan, Australia.
[6] Research School of Earth Sciences, Australian National University, Canberra, Australia.
[7] Centro Nacional de Investigación sobre la Evolución Humana (CENIEH), Burgos, Spain.
[8] School of Earth and Environmental Sciences, University of Queensland, Brisbane, Australia.
[9] Department of Histology and Embryology, School of Medicine, National and Kapodistrian University of Athens, Greece.
[10] Biomedical Research Foundation of the Academy of Athens, Greece.
[11] Faculty of Biology, Medicine and Health, University of Manchester, England.

© 2021, Kerns Verlag / https://doi.org/10.51315/9783935751377.002
Cite this article: Harvati, K., R. Grün, M. Duval, J. Zhao, A. Karakostis, V. Tourloukis, V. Gorgoulis, and M. Kouloukoussa. 2021. Direct U-series dating of the Apidima C human remains. In *Ancient Connections in Eurasia*, ed. by H. Reyes-Centeno and K. Harvati, pp. 37-55. Tübingen: Kerns Verlag. ISBN: 978-3-935751-37-7.

Apidima Cave C are potentially very significant in elucidating the arrival of the early Upper Paleolithic populations in Europe. Here we undertake direct Uranium-series dating of three human samples from Cave C, including the burial, to help clarify their chronology. Results suggest a minimum age of terminal Pleistocene for all three samples.

INTRODUCTION

Apidima is a cave complex situated on the coast of the Mani Peninsula, southern Greece, consisting of five caves (A-E) formed in the Upper Cretaceous–Late Eocene limestone of the coastal cliffs of the inner Mani (Fig. 1). The caves are situated very near the current sea level, Cave A being the lowermost at ca. 4 m above sea level (asl) and Caves C and D the highest (at ca. 19 m and 24 m asl, respectively). The caves were investigated by a team from the Museum of Anthropology of the Medical School of the National and Kapodistrian University of Athens between 1978 and 1985, and several important discoveries were made. These included two human fossil crania of Middle Pleistocene age from Cave A (Pitsios 1985, 1995, 1999; Harvati and Delson 1999; Harvati 2000; Harvati et al. 2009, 2011, 2019), considered among the most important paleoanthropological finds from southeast Europe. Their recent re-investigation showed the presence of an early modern human population, followed by a Neanderthal one, at the site in the Middle Pleistocene, and provided evidence of an early *Homo sapiens* dispersal out of Africa that was both earlier and geographically more widespread than previously thought (Harvati et al. 2019). However, a number of less known, but potentially very important human remains have also been recovered from Cave C. These include a burial hypothesized to be of early Upper Paleolithic age, a find that, if confirmed, would be unique in Greece (Pitsios 1985, 1995, 1999; Mompheratou and Pitsios 1995; Ligoni and Papagrigorakis 1995; Harvati et al. 2009; Tourloukis and Harvati 2018) and in Europe (d'Errico and Vanhaeren 2015).

Human remains reported from Cave C include a partial skeleton, as well as isolated dental remains and skeletal elements likely representing additional individuals (e.g., Mompheratou and Pitsios 1995; personal observation). The skeleton (LAO 1/S3, or Apidima 3) is represented by much of the postcranium, a mandibular fragment preserving the left molar series and possibly isolated teeth. It has been interpreted as a burial of a young woman. Sex was attributed on the basis of the pelvic morphology (Pitsios 1999), whereas age was estimated from dental attrition (Ligoni and Papagrigorakis 1995; Pitsios 1999). More than 40 (41 reported by Pitsios 1985, 43 by Pitsios 1999) pierced shells of *Nassa neritea* (Karali 1995) were reportedly recovered around the upper part of the skeleton (Pitsios 1999) and were considered to represent personal ornaments associated with the burial. A few lithic artifacts reportedly found together with this skeleton were tentatively assigned to the Aurignacian (Darlas 1995). Pitsios (1999) proposed a date of ca. 30 ka for this burial on the basis of his own stratigraphic observations, the tentative attribution of the lithics to the Aurignacian by Darlas (1995) and on ESR dates from cave sediments by Liritzis and Maniatis (1995).

The Upper Paleolithic is very rare in Greece and is known from only a handful of sites (e.g., Harvati et al. 2009; Harvati 2016; Tourloukis and Harvati 2018). Furthermore, human remains associated with the Aurignacian are very scarce throughout Europe, usually consisting of isolated specimens (most frequently teeth) rather than burials, even though a total absence of Aurignacian burials is not conclusive (Riel-Salvatore and Gravel-Miguel 2013; d'Errico and Vanhaerean 2015). Elaborate inhumations with ornaments, such as beads manufactured from shells, are overall scarce and usually more common in the middle and later parts of the Upper Paleolithic (Riel-Salvatore and Gravel-Miguel 2013). A possible early Upper Paleolithic chronology for Apidima C and the human remains found there is therefore of great interest. However, the age estimate proposed by Pitsios (1999) is largely conjectural. Pitsios (1999) does not specify how his stratigraphic observations can indicate a temporal range. The attribution of the lithics to the Aurignacian is tentative (Darlas 1995) and their association with the skeleton cannot be ascertained from the information published; with the exception of one specimen, a blade, the exact provenance of the lithics is either not specified or reported as probably unrelated to the context of the burial (Mompheratou

Fig. 1.
The Apidima Cave Complex, showing the position of the five caves, including Cave C. Inlet shows the geographic location of the Apidima, Kalamakia and Lakonis Paleolithic sites on the map of Mani Peninsula, Southern Peloponnese.

and Pitsios 1995: 37; but see also Darlas 1995: 59). Finally, while Liritzis and Maniatis (1995) produced two ESR dates of 20–30 ka and 25–45 ka for two travertine samples, these samples were taken from the opening of Cave D and B, respectively, and therefore have no bearing on either Cave C or the burial uncovered there (see also Harvati et al. 2009).

Here we conduct direct dating of the human remains from Apidima C, including the burial as well as two isolated teeth, using U-series dating in order to resolve this question. This effort was undertaken as part of the new research program of the Museum of Anthropology of the Medical School, National and Kapodistrian University of Athens, Greece, in collaboration with the Paleoanthropology group at the University of Tübingen, Germany, and the University of Bergen, Norway.

MATERIALS AND METHODS

Samples were selected from the Museum of Anthropology's collections of human remains excavated at Apidima C in the 1980s (Mompherratou and Pitsios 1995). LAO 1 S5 (Fig. 2A) is an isolated upper molar with extensive crown attrition (advanced stage 2, erosion across the entire dentine layer). It was found in the same context as the second specimen, LAO 1 S6 (Fig. 2B), a likely isolated premolar. Its extreme degree of attrition (stage 3, exposed pulp cavity; see Burns 2015) makes its exact anatomical allocation difficult. Two further specimens were selected from the bones associated with the burial of the female skeleton, Apidima 3: A fragment of the sternum (LAO 1 S3_18; Fig. 2C) and a fragment of a pelvic iliac bone (LAO 1 S3_12; Fig. 2D). Permission for sampling was obtained from the Ministry of Culture and Sports, Athens (ΥΠΠΟΑ/ΓΔΑΠΚ/ΔΣΑΝΜ/ΤΕΕ/Φ77/299995/215105/2663/281). All specimens were 3-d scanned before sampling using a handheld structured-light scanner with a maximum scanning accuracy of 50 microns, and high resolution casts were obtained of the two dental remains so as to create a complete record of their anatomy before the sampling procedure was undertaken. Of the four specimens, the iliac fragment (LAO 1 S3_12) did not preserve an appropriate cross-section for analysis and was therefore not used. The remaining samples were assigned the following laboratory reference numbers: LAO 1 S5 (isolated molar): 3776, LAO 1 S6 (isolated premolar): 3777, LAO 1 S3_12 (sternum fragment from female burial): 3778 (see Fig. 3A).

U-SERIES ANALYSIS

U-series dating is based on the different chemical behavior of uranium (U) and thorium (Th). While uranium is water solvable, thorium is not. As a result, minerals precipitated from water contain uranium isotopes (specifically ^{238}U, ^{234}U and ^{235}U). In a closed system, ^{234}U decays to ^{230}Th. The activity ratio of the two isotopes can be used to determine a U-series age. The ^{230}Th/^{234}U activity ratio starts with zero and grows

Fig. 2.
The human remains selected for sampling for dating analysis:
A. LAO 1 S5 (3776) isolated molar;
B. LAO 1 S6 (3777) isolated premolar;
C. LAO 1 S3_12 (3778) sternum fragment from female burial;
D. LAO 1 S3_12, iliac fragment from female burial, not used.

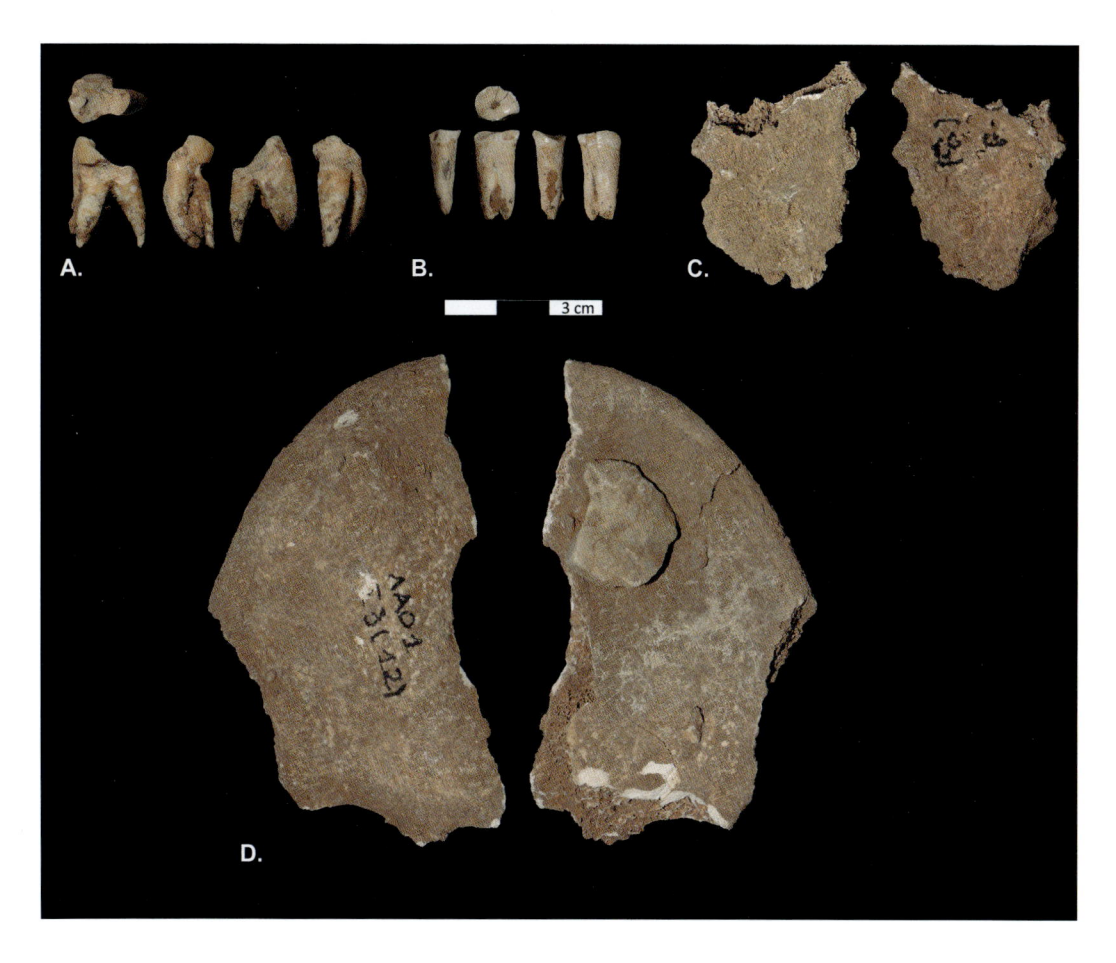

over time (about 600,000 years) into equilibrium when the ^{230}Th/^{234}U ratio is indistinguishable from unity. However, bones and teeth are not closed systems; they accumulate their uranium while they are buried in the ground. The actual U-uptake history can be highly complex (Grün et al. 2014), but generally leads to age calculations that underestimate the burial age of the specimen. It is possible to address the problem of the unknown U-uptake history with a variety of diffusion models (e.g., through the diffusion-adsorption model described by Pike et al. 2002, or the diffusion-adsorption-decay model of Sambridge et al. 2012). However, all these models are based on continuous diffusion processes and cannot recognize longer initial phases with no or little U-diffusion. This problem can be addressed in teeth by combining U-series and ESR methods (Grün et al. 1988). In the context of this study, ESR analysis was not feasible because of time constraints. To reiterate, the U-series ages reported here are apparent closed system age estimates, which most likely underestimate the burial ages of the specimens.

The U-series analyses were carried out using laser ablation, inductively coupled plasma multi-collector mass spectrometry (LA-ICP-

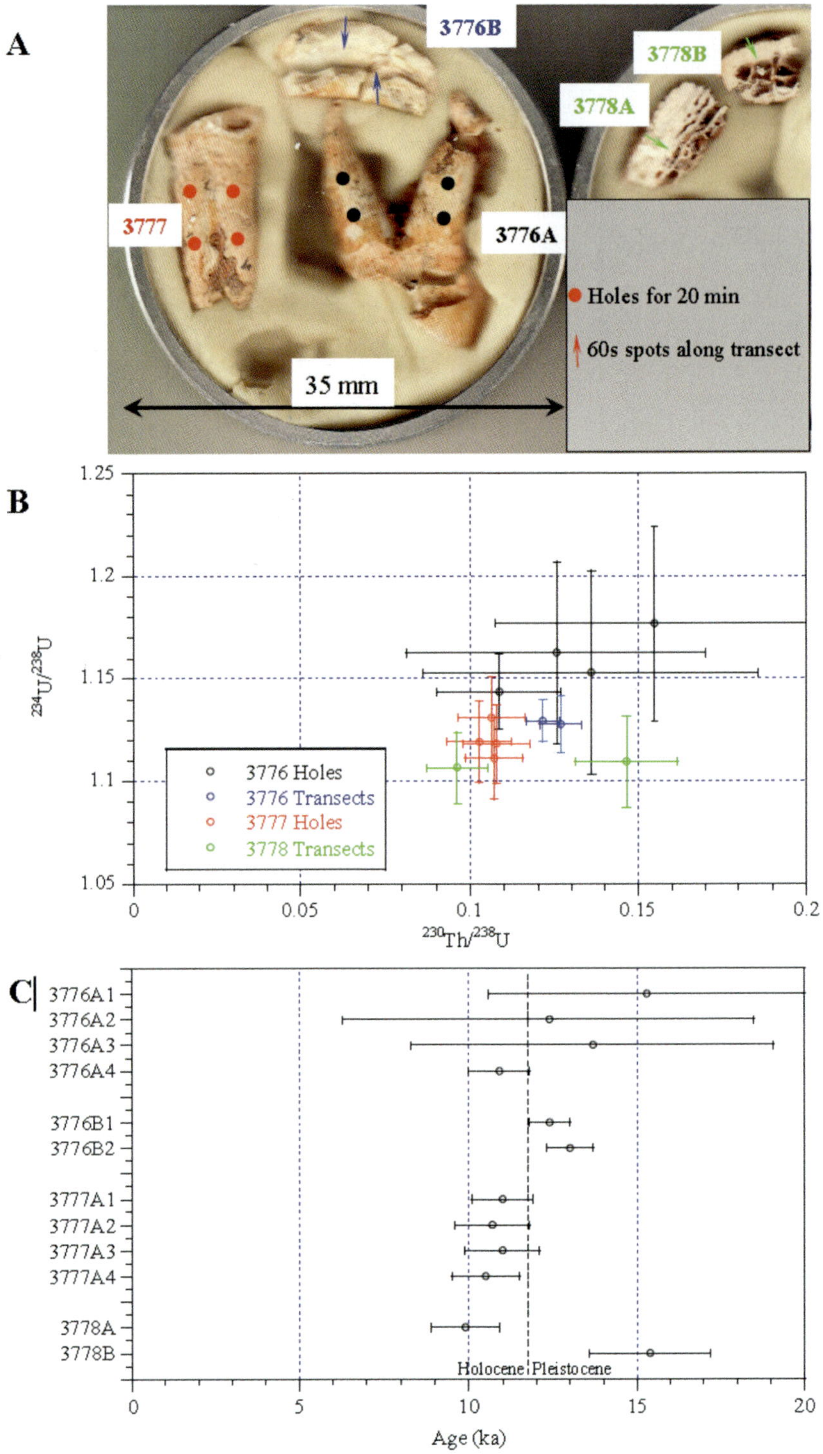

Fig. 3.
Samples and results of U-series analysis:
A. Samples and locations of the laser ablation analyses;
B. ^{230}Th/^{238}U vs ^{234}U/^{238}U activity ratios;
C. Age results with Pleistocene/Holocene boundary (Walker et al. 2018).

MCMS), which minimizes sample destruction of valuable human fossils (e.g., Groucutt et al. 2018). The analyses followed the procedures that were detailed by Grün et al. (2014). Two different analytical strategies were applied: analyzing spots (each for 60 s) along transects (3776B and 3778) and drilling holes with the laser in stationary position for 20 minutes (3776A and 3777), the latter procedure was applied to minimize sample damage (Benson et al. 2013). Sample 3776 was very fragile and a fragment of one of the roots split off. As a result, this individual tooth was analyzed by drilling four holes into the main part and two transects across the root fragment.

All isotope ratios in this paper are activity ratios with 2-σ errors. Ages were calculated with the Isoplot (Ludwig 2012).

RESULTS

The results of the individual spot analyses are shown in Table 1 and those of the holes in Table 2. The data in Table 2 were binned for 10 cycles (corresponding to approximately 10s ablation). As can be seen from Table 2, the analyses of the first three holes of sample 3776A1 to A3 are associated with large errors due to the low U-concentrations (< 2.3 ppm at the surface). The other holes (3776A4 and 3777-1 to 3777-4) had higher U-concentrations at the surface but the ablation efficiency rapidly decreased with measurement length so that only the data of the first 160 to 200 s were used for age calculations. Samples 3776 and 3777 have extremely high elemental U/Th ratios, indicating that there was no interference from detrital Th. The U/Th ratios for sample 3778 are somewhat higher, particularly for 3778B. All individual LA spots and holes return finite age results, indicating that the teeth have apparently not experienced uranium leaching. This could, however, be only confirmed by combining U-series with ESR data (Grün et al. 1988).

The ^{230}Th/^{238}U and ^{234}U/^{238}U are shown in Figure 3B. There seems to be an overall trend of slightly increasing ^{234}U/^{238}U ratios with increasing ^{230}Th/^{238}U ratios, but the large errors (due to low U-concentrations and young ages) prevent any meaningful interpretations. For sample 3776, the results of the transects and holes are compatible. The biggest difference is observed for sample 3778 where the two transects yielded distinctively different ^{230}Th/^{238}U results, and subsequently apparent ages. Sample 3778B has also distinctively higher U and Th concentrations, which may indicate some incorporation of detrital U and Th into the sample. However, corrections for detrital Th lead only to slightly younger results (by 0.2 ka). The age differences observed between samples 3778A and B could be simply due to some delayed U-uptake or some more complex processes that we cannot really address with the two transects. The distribution of the U-series ages does not indicate that U-leaching has occurred.

Because of the large associated errors, the apparent U-series ages of the three samples are overall statistically indistinguishable.

3776B1	U (ppm)	Th (ppb)	U/Th	$^{230}Th/^{238}U$	$^{230}Th/^{238}U$ error	$^{234}U/^{238}U$	$^{234}U/^{238}U$ error	initial $^{234}U/^{238}U$	$^{234}U/^{238}U_i$ error	Age (ka)	Age error (ka)
1	8.22	0.21	38477	0.1136	0.0119	11.332	0.0179	11.376	0.0184	11.5	1.3
2	8.47	n.d.	nd	0.1279	0.0116	11.277	0.0166	11.326	0.0172	13.1	1.3
3	8.42	0.04	212315	0.1223	0.0092	11.292	0.0185	11.338	0.0191	12.5	1.0
4	8.13	0.31	26172	0.1158	0.0116	11.347	0.0174	11.392	0.0180	11.7	1.3
5	8.47	0.92	9222	0.1227	0.0105	11.265	0.0182	11.310	0.0187	12.6	1.2
6	6.64	0.35	19111	0.1269	0.0121	11.260	0.0216	11.307	0.0223	13.0	1.3
AVERAGE VALUES											
	8.06±0.59	0.24±0.43		0.1214	0.0050	11.296	0.0104	11.342	0.0107	12.4	0.6

3776B2	U (ppm)	Th (ppb)	U/Th	$^{230}Th/^{238}U$	$^{230}Th/^{238}U$ error	$^{234}U/^{238}U$	$^{234}U/^{238}U$ error	initial $^{234}U/^{238}U$	$^{234}U/^{238}U_i$ error	Age (ka)	Age error (ka)
1	7.19	2.36	3042	0.1285	0.0115	11.208	0.0163	11.255	0.0169	13.3	1.3
2	7.23	1.62	4453	0.1310	0.0115	11.221	0.0223	11.269	0.0230	13.5	1.3
3	6.91	0.35	19522	0.1288	0.0142	11.393	0.0223	11.445	0.0231	13.1	1.5
4	5.94	0.66	8974	0.1124	0.0136	11.208	0.0342	11.248	0.0352	11.5	1.5
5	5.84	1.03	5668	0.1315	0.0148	11.373	0.0361	11.426	0.0373	13.4	1.7
AVERAGE VALUES											
	6.62±0.61	1.21±0.80		0.1267	0.0062	11.278	0.0137	11.326	0.0142	13.0	0.7

cont. ⟶

3778A	U (ppm)	Th (ppb)	U/Th	$^{230}Th/^{238}U$	$^{230}Th/^{238}U$ error	$^{234}U/^{238}U$	$^{234}U/^{238}U$ error	initial $^{234}U/^{238}U$	$^{234}U/^{238}U_i$ error	Age (ka)	Age error (ka)
1	3.86	15.07	256	0.0848	0.0172	11.015	0.0279	11.040	0.0286	8.7	1.9
2	2.97	9.40	316	0.0963	0.0212	10.855	0.0371	10.880	0.0381	10.1	2.4
3	2.87	23.05	125	0.1071	0.0211	10.931	0.0337	10.961	0.0346	11.2	2.4
4	3.42	6.74	507	0.1105	0.0188	11.414	0.0408	11.459	0.0419	11.1	2.0
5	4.03	8.18	492	0.0868	0.0183	11.074	0.0296	11.101	0.0302	8.9	2.0
AVERAGE VALUES											
	3.43±0.45	12.49±6.69		0.0962	0.0089	11.066	0.0173	11.096	0.0177	9.9	1.0

3778B	U (ppm)	Th (ppb)	U/Th	$^{230}Th/^{238}U$	$^{230}Th/^{238}U$ error	$^{234}U/^{238}U$	$^{234}U/^{238}U$ error	initial $^{234}U/^{238}U$	$^{234}U/^{238}U_i$ error	Age (ka)	Age error (ka)
1	4.21	78.20	54	0.1520	0.0270	11.075	0.0512	11.125	0.0534	16.1	3.2
2	5.03	20.59	244	0.1655	0.0384	10.950	0.0378	10.999	0.0396	17.9	4.5
3	4.26	19.63	217	0.1352	0.0319	11.158	0.0464	11.205	0.0481	14.1	3.6
4	4.71	46.33	102	0.1329	0.0213	11.193	0.0314	11.240	0.0325	13.8	2.4
AVERAGE VALUES											
	4.55±0.39	41.2±27.6	111	0.1463	0.0153	11.094	0.0223	11.143	0.0231	15.4	1.8

Table 1. (left and above)
U-series results on cross section 1 (indicated by arrows on Fig. 3A). All errors are 2-σ. n.d.: not determined; Th concentration below background. Ages calculated with Isoplot (Ludwig 2012). To avoid correlated errors, individual age errors do not include errors from standard. Average age errors result from the combination of the errors of the mean and the standard.

3776A1	U (ppm)	Th (ppb)	U/Th	$^{230}Th/^{238}U$	$^{230}Th/^{238}U$ error	$^{234}U/^{238}U$	$^{234}U/^{238}U$ error	initial $^{234}U/^{238}U$	$^{234}U/^{238}U_i$ error	Age (ka)	Age error (ka)
1	2.23	1.63	1364	0.1373	0.0816	11.750	0.0859	11.818	0.0888	13.5	8.5
2	1.91	n.d.	n.d.	0.1271	0.1096	11.736	0.0676	11.798	0.0698	12.5	11.4
3	1.76	0.58	3012	0.1153	0.0897	12.042	0.0927	12.106	0.0952	11.0	9.0
4	1.66	1.20	1381	0.1947	0.0769	11.490	0.0801	11.577	0.0841	20.2	8.9
5	1.42	n.d.	n.d.	0.1464	0.1398	11.766	0.1842	11.840	0.1907	14.5	14.9
6	1.21	n.d.	n.d.	0.2358	0.1555	12.578	0.1864	12.747	0.1961	22.5	16.8
7	1.02	n.d.	n.d.	0.1632	0.1495	10.907	0.2187	10.953	0.2289	17.7	18.0
AVERAGE VALUES											
				0.1547	0.0441	11.769	0.0474	11.847	0.0492	15.3	4.7

3776A2	U (ppm)	Th (ppb)	U/Th	$^{230}Th/^{238}U$	$^{230}Th/^{238}U$ error	$^{234}U/^{238}U$	$^{234}U/^{238}U$ error	initial $^{234}U/^{238}U$	$^{234}U/^{238}U_i$ error	Age (ka)	Age error (ka)
1	2.06	4.08	505	0.1833	0.1064	12.185	0.0811	12.297	0.0847	17.7	11.2
2	1.95	2.78	700	0.1174	0.0403	12.111	0.1160	12.178	0.1191	11.1	4.2
3	1.81	n.d.	n.d.	0.0802	0.0995	11.516	0.1025	11.550	0.1046	7.9	10.1
4	1.66	1.68	989	0.1395	0.0780	11.466	0.0753	11.526	0.0780	14.1	8.5
5	1.50	n.d.	n.d.	0.0809	0.1252	10.953	0.1054	10.976	0.1077	8.4	13.5
6	1.35	2.18	620	0.1894	0.1993	10.955	0.1449	11.012	0.1529	20.7	24.1
7	1.13	0.72	1556	0.1911	0.1070	11.757	0.2063	11.856	0.2159	19.3	12.4
8	1.04	1.33	781	0.0774	0.1285	12.244	0.1379	12.290	0.1404	7.1	12.2
9	0.84	n.d.	n.d.	0.0602	0.2926	11.055	0.2565	11.073	0.2607	6.1	30.6
10	0.74	n.d.	n.d.	0.1336	0.2599	12.220	0.2173	12.300	0.2243	12.6	26.0
AVERAGE VALUES											
				0.1255	0.0583	11.627	0.0443	11.685	0.0458	12.4	6.1

3776A3	U (ppm)	Th (ppb)	U/Th	^{230}Th/^{238}U	^{230}Th/^{238}U error	^{234}U/^{238}U	^{234}U/^{238}U error	initial ^{234}U/^{238}U	^{234}U/^{238}U$_i$ error	Age (ka)	Age error (ka)
1	2.03	May-88	346	0.2456	0.0762	13.168	0.1082	13.374	0.1135	22.4	7.9
2	1.91	0.52	3703	0.0946	0.1000	11.411	0.0958	11.449	0.0981	9.4	10.4
3	1.76	n.d.	n.d.	0.1289	0.0605	11.491	0.1512	11.547	0.1560	13.0	6.7
4	1.61	n.d.	n.d.	0.2235	0.1092	11.959	0.1102	12.087	0.1163	22.5	12.3
5	1.47	n.d.	n.d.	0.0610	0.0508	11.504	0.1358	11.529	0.1378	5.9	5.1
6	1.31	n.d.	n.d.	0.1031	0.1428	11.620	0.1181	11.667	0.1212	10.1	14.7
7	1.19	0.86	1388	0.1574	0.3060	10.665	0.1456	10.698	0.1526	17.4	36.7
8	1.06	n.d.	n.d.	0.2171	0.1442	11.192	0.1551	11.274	0.1645	23.5	17.7
9	0.91	n.d.	n.d.	0.1055	0.2293	12.340	0.2265	12.406	0.2320	9.7	22.1
10	0.79	n.d.	n.d.	0.1080	0.1365	10.990	0.2990	11.023	0.3078	11.3	15.3
AVERAGE VALUES											
				0.1359	0.0506	11.531	0.0497	11.591	0.0515	13.7	5.4

cont. →

Table 2. (left, above and the following 5 pages)
U-series results from continuous laser drilling (filled circles on Fig. 3A). Apparent U and Th concentrations partly indicate the declining ablation yield with depth 1. All errors are 2-σ. n.d.: not determined; Th concentration below background. Ages calculated with Isoplot (Ludwig 2012)). To avoid correlated errors, individual age errors do not include errors from standard. Average age errors result from the combination of the errors of the mean and the standard.

3776A4	U (ppm)	Th (ppb)	U/Th	^{230}Th/^{238}U	^{230}Th/^{238}U error	^{234}U/^{238}U	^{234}U/ error	initial ^{234}U/^{238}U	^{234}U/^{238}U$_i$ error	Age (ka)	Age error (ka)
1	7.76	4.24	1831	0.1269	0.0182	13.214	0.0583	13.316	0.0597	11.0	1.7
2	7.25	2.57	2821	0.1048	0.0143	11.522	0.0572	11.567	0.0586	10.4	1.6
3	6.75	1.35	5014	0.0856	0.0045	11.315	0.0307	11.347	0.0314	8.6	0.5
4	6.39	1.07	5955	0.1131	0.0241	11.569	0.0518	11.619	0.0532	11.2	2.6
5	5.83	0.18	32711	0.1294	0.0271	11.474	0.0383	11.529	0.0396	13.0	2.9
6	5.14	1.13	4533	0.1117	0.0427	11.674	0.0485	11.726	0.0498	11.0	4.4
7	4.89	0.72	6788	0.0938	0.0319	11.847	0.0584	11.895	0.0596	9.0	3.2
8	4.38	1.89	2317	0.1319	0.0282	11.440	0.0548	11.496	0.0566	13.3	3.1
9	3.79	1.05	3605	0.1008	0.0313	11.242	0.0680	11.279	0.0698	10.2	3.4
10	3.37	1.22	2762	0.1186	0.0517	11.280	0.0651	11.325	0.0671	12.1	5.6
11	2.99	n.d.	n.d.	0.1226	0.0534	11.101	0.0510	11.142	0.0527	12.8	5.9
12	2.74	n.d.	n.d.	0.1285	0.0743	11.010	0.0560	11.049	0.0580	13.5	8.3
13	2.56	0.06	41499	0.1416	0.0323	11.172	0.0658	11.222	0.0683	14.8	3.7
14	2.36	0.02	106695	0.0993	0.0371	11.119	0.0795	11.152	0.0816	10.2	4.1
15	2.21	0.65	3412	0.0974	0.0488	11.634	0.0834	11.679	0.0853	9.5	5.0
16	2.12	0.22	9616	0.1086	0.0582	11.242	0.1414	11.281	0.1454	11.1	6.4
17	1.94	n.d.	n.d.	0.0676	0.0319	10.944	0.1366	10.963	0.1391	7.0	3.5
18	1.80	1.52	1186	0.0972	0.0365	10.974	0.1145	11.002	0.1175	10.1	4.1
19	1.59	n.d.	n.d.	0.0399	0.0538	11.741	0.0630	11.759	0.0636	3.8	5.2
20	1.55	n.d.	n.d.	0.0910	0.0677	11.271	0.0597	11.304	0.0611	9.2	7.1
21	1.44	n.d.	n.d.	0.1308	0.0987	11.664	0.1297	11.726	0.1339	13.0	10.5
22	1.32	2.61	505	0.1255	0.0422	11.640	0.1551	11.698	0.1598	12.4	4.8
AVERAGE VALUES											
				0.1085	0.0088	11.438	0.0183	11.483	0.0188	10.9	0.9

3777-1	U (ppm)	Th (ppb)	U/Th	$^{230}Th/^{238}U$	$^{230}Th/^{238}U$ error	$^{234}U/^{238}U$	$^{234}U/^{238}U$ error	initial $^{234}U/^{238}U$	$^{234}U/^{238}U_i$ error	Age (ka)	Age error (ka)
1	8.36	17.20	486	0.1418	0.0225	11.084	0.0538	11.131	0.0559	14.9	2.6
2	7.73	5.40	1432	0.1319	0.0157	10.908	0.0614	10.945	0.0636	14.1	2.0
3	7.31	1.98	3687	0.1108	0.0265	11.343	0.0269	11.386	0.0277	11.2	2.8
4	6.93	0.83	8315	0.0949	0.0286	10.952	0.0848	10.979	0.0870	9.9	3.2
5	6.27	0.48	13099	0.0946	0.0289	11.254	0.0724	11.288	0.0741	9.6	3.1
6	5.61	2.55	2198	0.1298	0.0295	11.046	0.0486	11.087	0.0503	13.6	3.4
7	5.03	0.41	12155	0.1097	0.0198	11.230	0.0496	11.269	0.0511	11.2	2.2
8	4.37	3.50	1247	0.1057	0.0216	11.275	0.0714	11.314	0.0733	10.7	2.4
9	3.80	n.d.	n.d.	0.0959	0.0376	11.177	0.0364	11.210	0.0373	9.8	4.0
10	3.41	n.d.	n.d.	0.0854	0.0336	10.919	0.0615	10.942	0.0629	8.9	3.7
11	3.04	n.d.	n.d.	0.0990	0.0452	10.962	0.0639	10.991	0.0656	10.3	5.0
12	2.81	n.d.	n.d.	0.1124	0.0438	10.883	0.0974	10.913	0.1004	11.9	5.0
13	2.56	n.d.	n.d.	0.0825	0.0550	11.015	0.0631	11.039	0.0646	8.5	5.9
14	2.52	0.16	16029	0.0907	0.0517	11.197	0.0454	11.229	0.0465	9.2	5.5
15	2.36	n.d.	n.d.	0.0910	0.0291	11.497	0.0800	11.536	0.0818	9.0	3.1
16	2.17	0.80	2726	0.0585	0.0250	11.108	0.0986	11.127	0.1001	5.9	2.6
AVERAGE VALUES											
				0.1070	0.0085	11.113	0.0200	11.148	0.0206	11.0	0.9

cont. →

3777-2	U (ppm)	Th (ppb)	U/Th	$^{230}Th/^{238}U$	$^{230}Th/^{238}U$ error	$^{234}U/^{238}U$	$^{234}U/^{238}U$ error	initial $^{234}U/^{238}U$	$^{234}U/^{238}U_i$ error	Age (ka)	Age error (ka)
1	6.92	21.27	326	0.1216	0.0336	11.204	0.0538	11.247	0.0556	12.5	3.7
2	6.82	3.25	2101	0.1108	0.0358	11.271	0.0485	11.312	0.0499	11.3	3.9
3	6.56	0.90	7286	0.1134	0.0319	11.200	0.0504	11.240	0.0519	11.6	3.5
4	5.99	n.d.	n.d.	0.1029	0.0254	11.243	0.0581	11.280	0.0596	10.5	2.8
5	5.18	0.56	9178	0.1217	0.0387	11.033	0.0698	11.071	0.0722	12.7	4.4
6	4.47	1.86	2407	0.1068	0.0162	11.416	0.0536	11.459	0.0550	10.7	1.8
7	3.95	n.d.	n.d.	0.1103	0.0289	12.051	0.0583	12.112	0.0598	10.5	2.9
8	3.57	1.85	1933	0.0796	0.0334	10.823	0.0652	10.843	0.0666	8.3	3.7
9	3.21	n.d.	n.d.	0.1014	0.0454	11.537	0.0495	11.581	0.0507	10.0	4.7
10	2.93	0.92	3192	0.0801	0.0370	11.002	0.0711	11.025	0.0726	8.2	4.0
11	2.66	1.15	2314	0.1055	0.0633	12.124	0.0516	12.184	0.0529	9.9	6.2
12	2.55	0.94	2720	0.1253	0.0664	11.089	0.1051	11.130	0.1086	13.1	7.5
13	2.28	n.d.	n.d.	0.0790	0.0451	11.782	0.0823	11.820	0.0839	7.6	4.5
14	2.22	n.d.	n.d.	0.1195	0.0437	11.629	0.1207	11.684	0.1242	11.8	4.7
15	1.86	n.d.	n.d.	0.1435	0.0689	11.664	0.0819	11.733	0.0848	14.3	7.4
16	1.48	n.d.	n.d.	0.1071	0.0677	11.609	0.1092	11.657	0.1121	10.5	7.1
17	1.43	n.d.	n.d.	0.0506	0.0363	11.246	0.1048	11.264	0.1062	5.0	3.7
18	1.35	n.d.	n.d.	0.0730	0.0461	10.779	0.1136	10.796	0.1159	7.7	5.1
19	1.29	n.d.	n.d.	0.0770	0.0682	10.135	0.1451	10.139	0.1486	8.6	8.1
AVERAGE VALUES											
				0.1062	0.0098	11.311	0.0197	11.352	0.0202	10.7	1.1

3777-3	U (ppm)	Th (ppb)	U/Th	$^{230}Th/^{238}U$	$^{230}Th/^{238}U$ error	$^{234}U/^{238}U$	$^{234}U/^{238}U$ error	initial $^{234}U/^{238}U$	$^{234}U/^{238}U_i$ error	Age (ka)	Age error (ka)
1	7.03	18.27	385	0.1409	0.0228	11.139	0.0589	11.187	0.0611	14.7	2.7
2	6.73	5.32	1266	0.1133	0.0242	10.868	0.0260	10.898	0.0269	12.0	2.7
3	6.34	0.37	17266	0.1115	0.0291	11.167	0.0612	11.206	0.0630	11.5	3.2
4	5.76	1.80	3205	0.0868	0.0301	11.674	0.0512	11.714	0.0522	8.4	3.1
5	4.97	0.25	19934	0.0911	0.0300	11.283	0.0611	11.317	0.0625	9.2	3.2
6	4.38	n.d.	n.d.	0.1407	0.0318	11.151	0.0674	11.200	0.0699	14.7	3.7
7	3.86	0.92	4189	0.0961	0.0287	11.004	0.0605	11.032	0.0620	10.0	3.2
8	3.47	0.88	3933	0.0978	0.0418	11.101	0.0615	11.132	0.0631	10.1	4.5
9	3.13	n.d.	n.d.	0.1534	0.0460	11.518	0.0652	11.587	0.0677	15.6	5.1
10	2.72	0.59	4598	0.1165	0.0460	10.789	0.0808	10.817	0.0835	12.5	5.3
11	2.52	n.d.	n.d.	0.0989	0.0323	11.211	0.0569	11.246	0.0583	10.1	3.5
12	2.27	n.d.	n.d.	0.1034	0.0376	11.473	0.0761	11.517	0.0781	10.3	4.0
13	2.29	0.02	99403	0.0860	0.0669	11.888	0.0969	11.932	0.0989	8.2	6.6
14	2.11	n.d.	n.d.	0.0968	0.0310	11.033	0.0923	11.063	0.0947	10.0	3.5
15	1.83	2.06	884	0.1119	0.0436	10.643	0.0771	10.665	0.0796	12.1	5.1
16	1.62	1.46	1111	0.1028	0.0744	11.148	0.0836	11.183	0.0859	10.6	8.0
17	1.49	n.d.	n.d.	0.0615	0.0465	10.710	0.1119	10.723	0.1138	6.5	5.1
18	1.36	n.d.	n.d.	0.0566	0.0775	10.715	0.1135	10.727	0.1153	5.9	8.4
19	1.23	n.d.	n.d.	0.0748	0.0670	10.856	0.0995	10.875	0.1015	7.8	7.3
20	1.13	0.56	2024	0.0680	0.0953	12.158	0.1115	12.197	0.1133	6.3	9.1
AVERAGE VALUES											
				0.1077	0.0099	11.183	0.0190	11.220	0.0196	11.0	1.1

cont. ⟶

3777-4	U (ppm)	Th (ppb)	U/Th	^{230}Th/^{238}U	^{230}Th/^{238}U error	^{234}U/^{238}U	^{234}U/^{238}U error	initial ^{234}U/^{238}U	^{234}U/^{238}U$_i$ error	Age (ka)	Age error (ka)
1	7.12	15.09	472	0.1312	0.0277	11.337	0.0351	11.388	0.0363	13.4	3.0
2	6.77	0.29	23626	0.0971	0.0263	11.018	0.0503	11.048	0.0517	10.1	2.9
3	6.26	n.d.	n.d.	0.1106	0.0264	11.200	0.0445	11.239	0.0458	11.3	2.9
4	5.94	0.32	18482	0.1009	0.0245	10.729	0.0661	10.751	0.0680	10.8	2.8
5	5.52	0.48	11405	0.1074	0.0226	11.011	0.0468	11.044	0.0482	11.2	2.5
6	4.72	0.74	6383	0.1021	0.0374	11.339	0.0829	11.379	0.0850	10.3	4.0
7	4.21	n.d.	n.d.	0.0931	0.0344	11.326	0.0922	11.362	0.0944	9.4	3.7
8	3.64	n.d.	n.d.	0.0891	0.0290	11.651	0.0738	11.692	0.0754	8.7	3.0
9	3.30	n.d.	n.d.	0.0883	0.0664	11.386	0.0680	11.421	0.0695	8.8	6.9
10	3.02	0.10	30743	0.0995	0.0371	11.942	0.0641	11.995	0.0655	9.5	3.7
11	2.91	n.d.	n.d.	0.0957	0.0361	10.737	0.0627	10.758	0.0644	10.2	4.1
12	2.82	n.d.	n.d.	0.0750	0.0253	10.843	0.0602	10.862	0.0615	7.8	2.8
13	2.53	n.d.	n.d.	0.1015	0.0389	11.164	0.0875	11.198	0.0898	10.4	4.3
14	2.32	n.d.	n.d.	0.1270	0.0568	10.786	0.0622	10.817	0.0645	13.7l	6.6
15	2.13	0.68	3122	0.0818	0.0507	11.859	0.0900	11.901	0.0917	7.8	5.0
16	1.90	n.d.	n.d.	0.0755	0.0740	11.032	0.0748	11.055	0.0764	7.7	7.9
17	1.81	n.d.	n.d.	0.1289	0.0406	10.514	0.1326	10.535	0.1377	14.3	5.2
18	1.64	n.d.	n.d.	0.1710	0.0679	0.9917	0.0782	0.9912	0.0829	20.7	9.2
AVERAGE VALUES											
				0.1026	0.0094	11.193	0.0199	11.229	0.0205	10.5	1.0

DISCUSSION AND CONCLUSIONS

Our age results imply that all samples have experienced a U-uptake event that corresponds to the Pleistocene/Holocene transition at 11.7 ka b2k (before the year 2000; Walker et al. 2018). Considering that the samples were most likely an open system for some time, particularly in view of the young apparent ages, it can be reasonably envisaged that the U-uptake took place during the terminal Pleistocene. However, it is important to consider that our dates represent minimum age constraints for the samples, and therefore do not exclude an earlier Upper Paleolithic age for the human remains and burial. Nevertheless, in light of these findings, the association of the Apidima 3 skeleton with the Aurignacian lithics, as well as the attribution of the lithics to the Aurignacian, should be re-evaluated. A better understanding of the chronology of the human samples would be gained by ESR analyses on samples 3776 and 3777 (e.g., Brumm et al. 2016). Unfortunately, radiocarbon dating of these samples (e.g., Higham at al. 2014) was not possible due to poor collagen preservation (Higham pers. comm.). However, it might be possible to apply that method to the pierced shell remains associated with Apidima 3 (Douka 2017). Finally, the question of association of the human remains with the material cultural remains recovered at the site can only be resolved through renewed fieldwork and excavation aiming to resolve the stratigraphy, depositional context and site formation processes at Apidima C. In summary, our results confirm a Pleistocene age for the Apidima C human remains, but further research is necessary to assess their association with the early Upper Paleolithic assemblages described for this site.

ACKNOWLEDGMENTS

The authors would like to thank Yuexing Feng, University of Queensland, for his invaluable help with the Laser Ablation ICP-MS measurements and Nick Thompson for his help with photographing the specimens used in our analysis. M. Duval's research is funded by the Australian Research Council Future Fellowship FT150100215 and the Spanish Ramón y Cajal Fellowship RYC2018-025221-I. This research was supported by the European Research Council ERC CoG CROSSROADS (724703). We are grateful to the Greek Ministry of Culture and Sports for their support and to the anonymous reviewer who greatly helped improve this manuscript.

REFERENCES

Benson, A., L. Kinsley, A. Defleur, H. Kokkonen, M. Mussi, and R. Grün. 2013. Laser ablation depth profiling of U-series and Sr isotopes in human fossils. *Journal of Archaeological Science* 40: 2991–3000.

Brumm, A., G. van den Bergh, M. Storey, I. Kurniawan, B. V. Alloway, R. Setiawan, E. Setiyabudi, R. Grün, M. W. Moore, D. Yurnaldi, M. R. Puspaningrum, U. P. Wibowo, H. Insani, I. Sutisna, J. A. Westgate, N. J. G. Pearce, M. Duval, H. J. M. Meijer, F. Aziz, T. Sutikna, S. van der Kaars, and M. J. Morwood. 2016. Age and context of the oldest known hominin fossils from Flores. *Nature* 534: 249–253.

Burns, R. K. 2015. *Forensic anthropology training manual.* Routdlege: New York.

Darlas, A. 1995. Τα λίθινα εργαλεία του σκελετού ΛΑΟ 1/Σ 3 (Απήδημα – Μάνη). *Acta Anthropologica* 1: 59–62.

d'Errico, F., and M. Vanhaeren. 2015. Upper Palaeolithic mortuary practices: Reflection of ethnic affiliation, social complexity, and cultural turnover. In *Death Rituals, Social Order and the Archaeology of Immortality in the Ancient World: 'Death Shall Have No Dominion'*, ed. by C. Renfrew, M. Boyd and I. Morley, pp. 45–62. Cambridge: Cambridge University Press.

Douka, K. 2017. Radiocarbon dating of marine and terrestrial shell. In Molluscs in Archaeology: methods, approaches and applications, ed. by M. J. Allen, pp. 381–389. Studying Scientific Archaeology 3. Oxbow Books.

Groucutt, H. S., R. Grün, I. S.A. Zalmout, N. A. Drake, S. J. Armitage, I. Candy, R. Clark-Wilson, J. Louys, P. S. Breeze, M. Duval, L. T. Buck, et al. 2018. Homo sapiens in Arabia by 85,000 years ago. *Nature Ecology & Evolution* 2: 800–809. DOI: 10.1038/s41559-018-0518-2.

Grün, R., S. Eggins, L. Kinsley, H. Mosely, and M. Sambridge. 2014. Laser ablation U-series analysis of fossil bones and teeth. *Palaeogeography, Palaeoclimatology, Palaeoecology* 416: 150–167.

Grün, R., H. P. Schwarcz, and J. M. Chadam. 1988. ESR dating of tooth enamel: Coupled correction for U-uptake and U-series disequilibrium. *Nuclear Tracks and Radiation Measurements* 14: 237–241.

Harvati, K. 2000. Apidima. In *Encyclopedia of Human Evolution and Prehistory*, ed. by E. Delson, I. Tattersall, J. A. van Couvering and A. S. Brooks, pp. 141–2. 2nd edition. New York: Garland Publishing.

Harvati, K. 2016. Paleoanthropology in Greece: Recent findings and interpretations. In *Paleoanthropology of the Balkans and Anatolia: Human Evolution and its Context*, ed. by K. Harvati and M. Roksandic, pp. 3–14. Vertebrate Paleobiology and Paleoanthropology Series. Dordrecht: Springer.

Harvati, K., and E. Delson. 1999. Conference report: Paleoanthropology of the Mani Peninsula (Greece). *Journal of Human Evolution* 36: 343–348.

Harvati, K., E. Panagopoulou, and C. Runnels. 2009. The paleoanthropology of Greece. *Evolutionary Anthropology* 18: 131–43.

Harvati, K., C. Stringer, and P. Karkanas. 2011. Multivariate analysis and classification of the Apidima 2 cranium from Mani, Southern Greece. *Journal of Human Evolution* 60: 246–250.

Harvati, K., C. Röding, A. Bosman, F. A. Karakostis, C. Stringer, P. Karkanas, N. Thompson, V. Koutoulidis, L. A. Moutopoulos, V. Gorgoullis, and M. Kouloukoussa. 2019. Apidima cave fossils provide earliest evidence of Homo sapiens in Eurasia. *Nature* 571: 502–504.

Higham, T., K. Douka, R. Wood, C. Bronk Ramsey, F. Brock, L. Basell, M. Marta Camps, A. Arrizabalaga, J. Baena, C. Barroso-Ruiz, C. Bergman, C. Boitard, P. Boscato, M. Caparros, N. J. Conard, C. Draily, A. Froment, B. Galvan, P. Paolo Gambassini, A. Garcia-Moreno, S. Grimaldi, P. Haesaerts, B. Holt, M. J. Iriarte-Chiapusso, A. Jelinek, J. F. Jorda Pardo, J. M. Maillo-Fernandez, A. Marom, J. Maroto, M. Menendez, L. Metz, E. Morin, A. Moroni, F. Negrino, E. Panagopoulou, M. Peresani, S. Pirson, M. de la Rasilla, J. Riel-Salvatore, A. Ronchitelli, D. Santamaria, P. Semal, L. Slimak, J. Soler, N. Soler, A. Villaluenga, R. Pinhasi, and R. Jacobi. 2014. The timing and spatiotemporal patterning of Neanderthal disappearance. *Nature* 512: 306–309.

Karali, L. 1995. Preliminary report on malacological material found in Apidima cave in Laconia. *Acta Anthropologica* 1: 159–162.

Ligoni, E., and M. Papagrigorakis. 1995. Odontological examination of skeleton LAO 1 / S3. *Acta Anthropologica* 1: 53–58.

Liritzis, Y., and Y. Maniatis. 1995. ESR experiments on calcites and bones for dating purposes. *Acta Anthropologica* 1: 65–98.

Ludwig, K. R. 2012. Isoplot. Excel package version 3.76.12.02.24 http://www.bgc.org/isoplot_etc/isoplot.html

Mompheratou, E., and T. K. Pitsios. 1995. Apidima, cave Gamma (Γ): The burial of female skeleton LAO 1 /S3. *Acta Anthropologica* 1: 27–51.

Pike, A. W. G., R. E. M. Hedges, and P. Van Calsteren. 2002. U-series dating of bone using the diffusion-adsorption model. *Geochimica et Cosmochimica Acta* 66: 4273–4286.

Pitsios, T. K. 1985. Παλαιοανθρωπολογικές έρευνες στη θέση «Απήδημα» της Μέσα Μάνης. Αρχαιολογία 15: 26–33.

Pitsios, T. K. 1995. Paleoanthropological Research at the cave site of Apidima, Laconia, Greece. *Acta Anthropologica* 1: 1–180.

Pitsios, T. 1999. Paleoanthropological research at the cave site of Apidima and the surrounding region (South Peloponnese, Greece). *Anthropologischer Anzeiger* 57: 1–11.

Riel-Salvatore, J., and C. Gravel-Miguel. 2013. Upper Palaeolithic mortuary practices in Eurasia: A critical look at the burial record. In *The Oxford handbook of the archaeology of death and burial*, ed. by L. Nilsson Stutz and S. Tarlow, pp. 303–46. Oxford: Oxford University Press.

Sambridge, M., R. Grün, and S. Eggins. 2012. U-series dating of bone in an open system: The diffusion-adsorption-decay model. *Quaternary Geochronology* 9: 42–53.

Tourloukis, V., and K. Harvati. 2018. The Palaeolithic record of Greece: A synthesis of the evidence and a research agenda for the future. *Quaternary International* 466: 48–65.

Walker, M., M. J. Head, M. Berkelhammer, S. Björck, H. Cheng, L. Cwynar, D. Fisher, V. Gkinis, A. J. Long., J. Lowe, R. Newnham, S. O. Rasmussen, and H. Weiss. 2018. Formal ratification of the subdivision of the Holocene Series/Epoch (Quaternary System/Period): Two new Global Boundary Stratotype Sections and Points (GSSPs) and three new stage/subseries. *Episodes* 41: 213–223.

CHAPTER 3

Profile orientation change through time in Upper Paleolithic parietal art

Matteo Scardovelli[1]

Abstract

Art traditions reflect beliefs, practices, customs and unconscious values. If it is relatively easy to study these features in historic art traditions, the same is not true for prehistoric times. In this case, archaeology has to rely on multidisciplinary approaches in order to counterbalance the deficiencies in documentation. The large quantity of images known to us from the Upper Paleolithic permits a quantitative study of different iconographic features and their change through time. With the program SPSS, I created a database of more than 2,000 images belonging to Upper Paleolithic parietal art, allowing me to study correlations among different discrete variables that open the possibility for finding specific cultural "markers." This represents an alternative to the study of stylistic features, a qualitative and valid methodology that is sometimes criticized for its lack of objectivity. In this paper, I will discuss the change through time in the preferential profile orientation of figures/images in order to confirm or question the cultural boundaries in Upper Paleolithic Europe.

INTRODUCTION

The aim of the present contribution is to study parietal art with the help of descriptive statistics in order to shed light on the chrono-cultural boundaries in Upper Paleolithic Europe. In my PhD thesis (Scardovelli 2017) I developed a methodology joining together archaeological research, statistical analyses and a semiotic approach inspired by embodied cognition theories. The statistical analyses, tables and graphs were developed from a database created with the program SPSS and containing more than 2,000 images from six major painted caves (Lascaux,

---

[1] Valcamonica Centre of Prehistoric Studies, Italy.

© 2021, Kerns Verlag / https://doi.org/10.51315/9783935751377.003
Cite this article: Scardovelli, M. 2021. Profile orientation change through time in Upper Paleolithic parietal art. In *Ancient Connections in Eurasia*, ed. by H. Reyes-Centeno and K. Harvati, pp. 57-72. Tübingen: Kerns Verlag. ISBN: 978-3-935751-37-7.

Chauvet, Rouffignac, Cosquer, Niaux, Pech-Merle) in addition to 24 minor caves containing less than 40 images each (see Table 1). The time-span under consideration extends from the Aurignacian to the Magdalenian (broadly, ~40,000 BP to ~12,000 BP), and all the caves are from central/southern France and the Cantabrian region in Spain. Table 1 indicates the exact number of images recorded for each cave divided into the four chrono-cultural units of the European Upper Paleolithic (Aurignacian, Gravettian, Solutrean, Magdalenian). Each chrono-cultural group is especially associated with one or two cave(s): most of the figures (~90%) from the Aurignacian period are found at Chauvet, while the Gravettian is represented mostly by the Quercy caves (Pech-Merle, Cougnac, ~50 and ~15%, respectively). Lascaux (~80%) and Cosquer (~17%) exemplify the Solutrean (for a discussion on the Solutrean attribution of Lascaux, see Scardovelli 2017: Sect. 3.4.2). The Magdalenian is mainly represented by Rouffignac (~60%) and Niaux (~27%). The existence of largely decorated caves especially associated with particular chrono-cultural groups will be addressed in the hypothesis.

Among the different features that can be analyzed statistically, I chose to consider the orientation of animal figures, that is, the direction, left or right, that the represented animal is facing in relation to the observer (for an example, see Fig. 1). The choice of this topic is based on the assumption that it is possible to draw general conclusions by analyzing if some groups of images show preferential left or right profile (for similar studies in art history, see Chateau 1965; Darras 1996: Chap. 12; Guillaumin 1961; Hinz 1974; Simounet 1975; Zazzo 1950; McManus and Humphrey 1973; Uhrbrock 1973; Sauvet 2005; Barrière 1997: 518). In this text, I investigate an important iconographic change in figure orientation through the Upper Paleolithic in Europe. I will present the quantitative analysis and I will provide a possible interpretation of this shift in profile choice. I would like to clarify that the two things are separated: the statistical demonstration that a significant change has occurred in profile preference during Upper Paleolithic parietal art is one thing; the interpretation of this shift is another thing. If the first data can be con-

Fig. 1.
The representation of a bison in Altamira Cave. The right bison is the original one, the left bison is a Photoshop transformation of the original image (https://commons.wikimedia.org/wiki/File:AltamiraBison.jpg).

Four major Upper Paleolithic cultures in Europe (BP)		Frequency	Percent
Aurignacian (34-29,000)	Les Fieux	2	0.5
	Chauvet	398	90.9
	Pair-non-Pair	28	6.4
	El Pindal	5	1.1
	Llonin	5	1.1
	Total	**438**	**100%**
Gravettian (30-20,000)	Pech-Merle	93	49.5
	Marcenac	11	5.9
	Cougnac	29	15.4
	Grotte des Merveilles	12	6.4
	Combe-Nègre I	3	1.6
	Travers de Janoye (Penne-Tarn)	5	2.7
	Mayrière Supérieure	2	1.1
	Cantal	1	0.5
	Cuzoul des Brasconies	1	0.5
	Escabasses	5	2.7
	Foissac	3	1.6
	Cuzoul de Mélanie	1	0.5
	Chauvet	8	4.3
	Cosquer	14	7.4
	Total	**188**	**100%**
Solutrean (22-17,000)	La Grèze	3	0.2
	Cosquer	218	17.3
	Lascaux	1008	79.9
	El Pindal	22	1.7
	Llonin	10	0.8
	Total	**1261**	**100%**
Magdalenian (17-12,000)	Sainte-Eulalie	36	8.6
	Moulin	4	1.0
	Christian (Lot)	2	0.5
	Bigourdane	4	1.0
	Abri Murat	2	0.5
	Réseau Clastres (Niaux)	5	1.2
	Niaux	114	27.2
	Rouffignac	248	59.2
	El Pindal	4	1.0
	Total	**419**	**100%**
unknown	Les Fieux	3	30.0
	Papetier	1	10.0
	La Loja	6	60.0
	Total	**10**	**100%**
	Total	**2316**	

Table 1.
The number of images recorded for each cave included in each chrono-cultural unit.

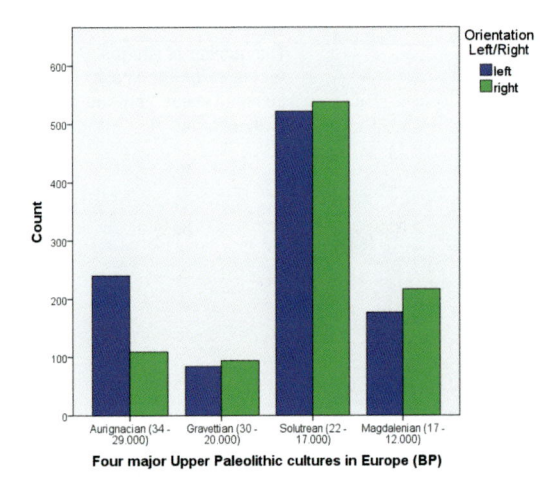

Fig. 2.
In this graph, it is possible to see the change in orientation preference through the Upper Paleolithic in Europe. During the Aurignacian there is a left profile preference and since the Gravettian the preference switches to right profile (chi$^2_{(3)}$ = 51,959, p < .001).

sidered as hardly questionable, the second one, belonging to the sphere of human interpretation, can always be debated and improved, as is true for any interpretation in any field of study.

I will now present the data concerning the orientation of the parietal art figures belonging to the database I created. Among the 2,316 figures present in this database, only 1,991 have a known or clear orientation (some figures are partially destroyed and unreadable, for others there is no profile, such as for figures facing the observer, and for a third group it was not possible to collect such information). In total, 1,033 of 1,991 figures show a preference for left orientation, representing the ~52 % of the total.

What is particularly interesting for determining chrono-cultural boundaries in Upper Paleolithic Europe is the diachronic change of this preference among the Aurignacian, the Gravettian, the Solutrean and the Magdalenian. In fact, as we can see in Figure 2, there has been a significant shift from left to right profile preference from the Aurignacian period to the following ones. At a descriptive level, the same trend seems to persist irrespective of the caves considered: in Figure 3 we can see that the Aurignacian is dominated by left-facing figures, as seen in Chauvet or other minor sites. The Gravettian and the Solutrean reveal more balanced results, with the only exception being the Quercy caves (Pech-Merle and Cougnac). Finally, the Magdalenian presents a large number of right-facing figures, with the only exception being some minor caves. In sum, we seem to be acknowledging a major shift in some aesthetic value of images in the passage from the Aurignacian to later periods. For correctly interpreting such a result, it is imperative to take a step back and consider the studies concerning figure orientation in general.

CONTEXTUALIZING THE RESEARCH: THE STUDY OF FIGURE ORIENTATION IN ART HISTORY

Several researchers (see above section) have tried to analyze different art traditions in the light of figure orientation, even if a global vision on this topic is still lacking. Broadly, it is possible to affirm that four kinds of influences, apart from chance, determine the orientation of a figure, and they can be labeled as follows: cultural, bio-mechanical, contextual and neurological. Their mutual interaction is debatable; therefore, it is preferable to study each influence individually. In this brief review, I will assume that the creator of images is a right-handed individual, and this for two main reasons: first, because humans have been preferentially right-handed for many millennia (Cashmore, Uomini, and Chapelain 2008; Frayer et al. 2012; Toth 1985); second, because some interesting

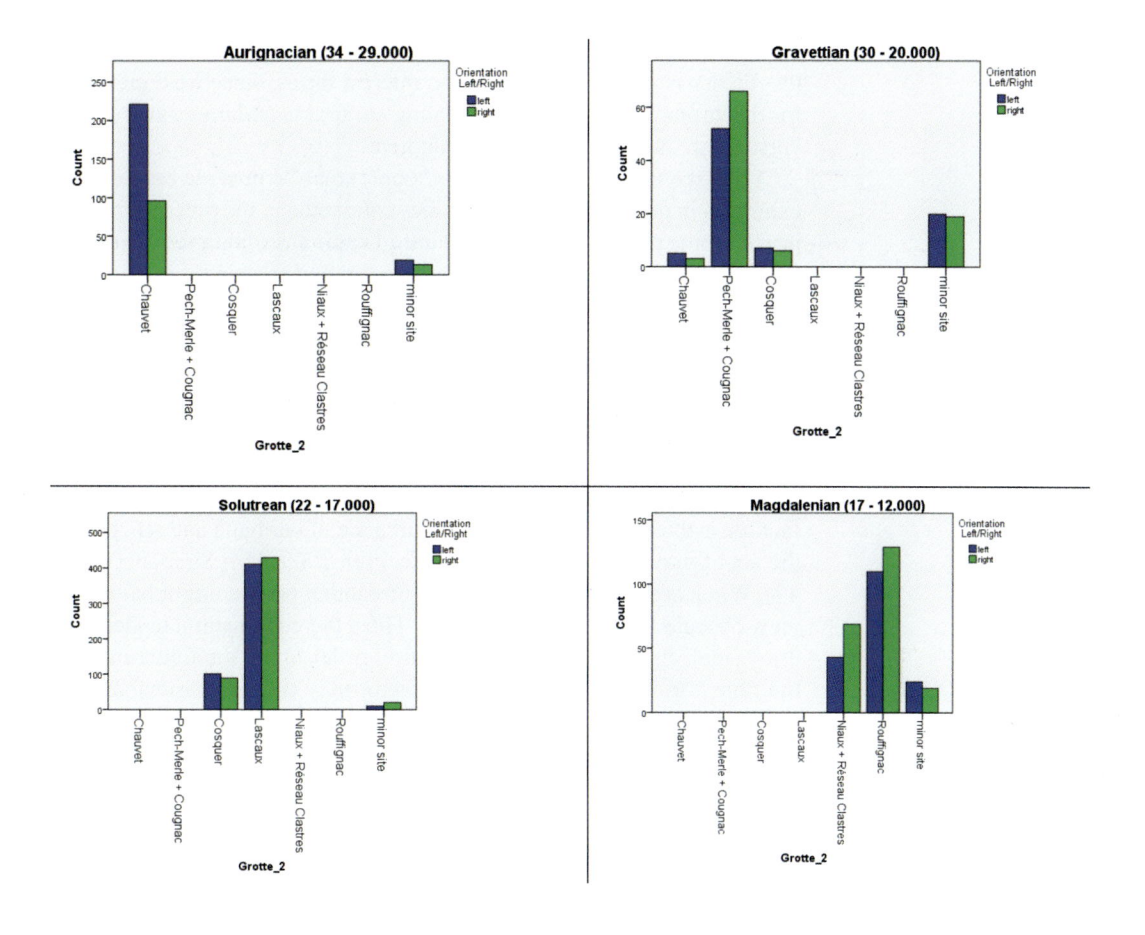

Fig. 3.
In this graph we observe the number of left-facing and right-facing figures in each cave divided into the four Upper Paleolithic chrono-cultural units.

observations seem to suggest that European paleo-artists were effectively right-handed (Groenen 1997).

The first sort of influence is cultural, and based on any sort of graphic convention. The more common cultural influence is the script habit of writing from left to right or from right to left, something that can strongly influence image profile orientation (Tosun and Vaid 2014). For Westerners who write from left to right, it is more natural to draw animals oriented to the left. During the Paleolithic there was no writing system; nonetheless, some cultural code related to figure orientation cannot be discounted (Sauvet 2005). But it is not my intention to take this kind of influence into account here, not least due to a clear lack of knowledge in this regard.

The second group of factors concerns the so-called "bio-mechanical" constraints. Some experts think that, for right-handers, the movement of the arm from left to right is more "natural" than the opposite movement (Van Sommers 1984: 91). As a consequence, there would be a generic preference for left-facing figures. According to the theory, this factor is strongly hard-wired, somehow applying to all artistic traditions. Whether this theory is valid or not, it is of no interest for us, since we are looking for diachronic transformation revealing cultural boundaries: bodily "universals" are not a good starting point here.

I call the third group of factors "contextual." I provide below some examples in order to show how studies conducted in the present time can predict some peculiarity of parietal art. In some cases, the orientation of a figure is determined by factors that go beyond the single image. One example is when an artist creates a composition, for example the scene of the Nativity in a Christian context. In a classic Nativity scene, all people in the painting look towards the newborn Jesus, so their place in the composition determines the direction of the face. Scenes are rare in Paleolithic art, and the difficulties in interpreting such scenes make the study of this factor highly complicated.

Another case, always falling into the general sphere of "contextual" factors, is the effect of the visual disturbance of our hand and arm during the act of drawing (Braswell and Rosengren 2000: 154; Simounet 1975: 49). When creating small figures, right-handed people might have their view obscured by their drawing hand. Thus, there is a natural tendency to draw small figures oriented to the left in order to see the figure most of the time during its realization. This factor may concern parietal art in a special way, since illumination in caves at that time was scarce. In fact, it is true that small images are oriented to the left more often than larger images (see Fig. 4) The significant relation between these two variables is also statistically validated (see graphic's legend). The trend seems to remain unchanged during the entire Upper Paleolithic (Fig. 5). But this time the statistical significance remains valid only for one chrono-cultural unit, i.e., the Solutrean, reflecting the fact that only Lascaux has a significant relation to the variable of profile orientation ($Chi^2_{(2)}$ = 10,370, p = 0.006). Despite the fact that only Lascaux shows a statistical significance for our purposes here, if we merely look at percentages, all caves show the same trend of leftward-facing small figures and rightward-facing larger figures (data not shown), with the only exception being the small caves considered altogether (but this might be simply a consequence of a lack of data, since in this category I only counted 122 figures). If the topic of this study was the phenomenon described in the present paragraph, I might of course deepen the statistical investigation. But the shift of orientation in respect to the dimension of the image is not my concern here; my goal is only to show how present-day studies can be used for interpreting some minor characteristics of parietal art.

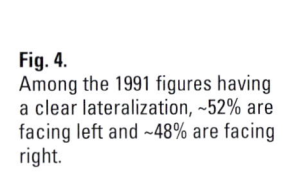

Fig. 4.
Among the 1991 figures having a clear lateralization, ~52% are facing left and ~48% are facing right.

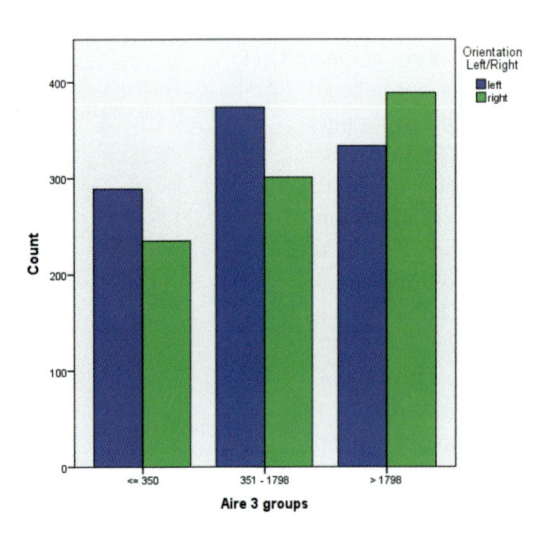

The fourth and last group of factors can be labeled as "neurological," and relate to brain hemispheric asymmetries (for a review, see Scardovelli 2017: Sec. 8.1). Before continuing, it is important to clarify something: the neurological aspects that I will consider have been studied in "passive" situations, when subjects were feeling something or looking at something. Nonetheless, it is possible to imply that the neurological aspects discussed below are analogous to the ones implied in "active" situations, such as during the act of drawing. As I have discussed in Scardovelli (2017: Sec. 8.1), following research of embodied cognition, the neurological processes involved in perception are often analogous to the neurological processes operating during action and ideation (for a general discussion on the topic of embodied cognition, see Barsalou 2008).

The two hemispheres differ in several characteristics, but here I will discuss only two aspects in particular: the first one is the Face Perception

Fig. 5.
In this graph, it is possible to see the change in orientation preference through the Upper Paleolithic in Europe. From the Aurignacian significant preference is for left profile; from the Gravettian on the preference switches to right profile.

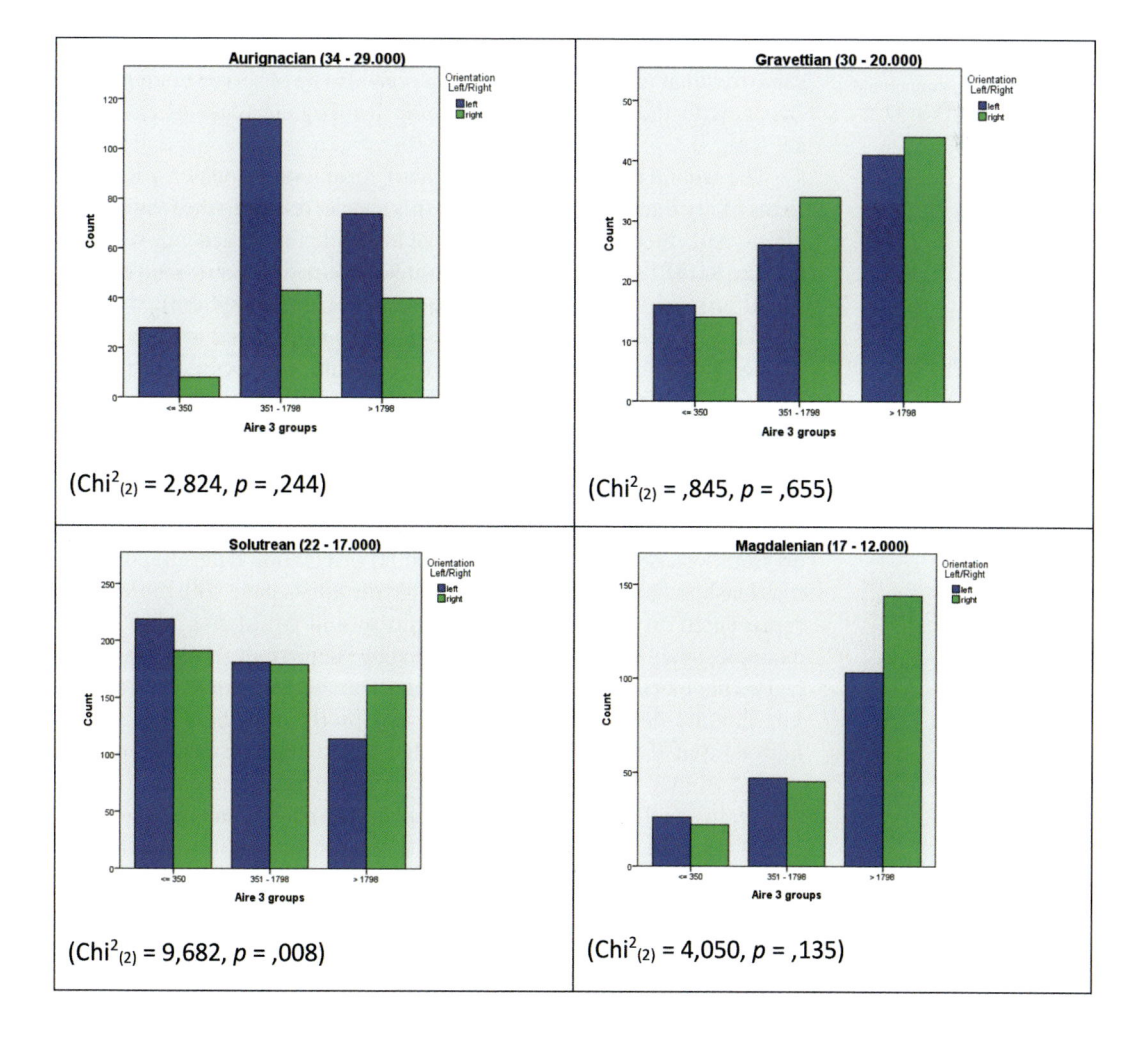

$(Chi^2_{(2)} = 2,824, p = ,244)$

$(Chi^2_{(2)} = ,845, p = ,655)$

$(Chi^2_{(2)} = 9,682, p = ,008)$

$(Chi^2_{(2)} = 4,050, p = ,135)$

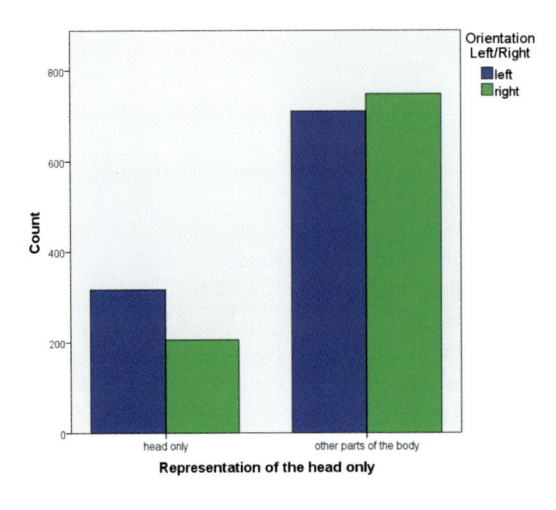

Fig. 6.
Graphic showing the change in the choice of the wall of the cave, left of right, through the four major Upper Paleolithic cultures in Europe.

Module, that is more commonly located in the right hemisphere (Gazzaniga and Smylie 1983). This determines facilitation, for our neurological system, in visualizing faces located in the left side of our visual field, and therefore faces looking left. This factor too can be detected in prehistoric art: representations of animal heads (without the body) show a significant preference for left orientation (see Fig. 6). The trend seems to remain unchanged throughout the entire Upper Paleolithic (Fig. 2). But, as was the case for the study of profile preference in relation to the dimensions of figures, here as well the trend is corroborated only by simple percentages and not by statistical significance, again with the only exception being the Solutrean period, dominated by Lascaux. The results concerning the Aurignacian and the Gravettian show a trend towards statistically significant results. The same trend, in mere percentage terms, can also be observed among single caves, with the only exception being Rouffignac (data of caves not shown).

The second neurological factor that I consider here, which will be the focus of my interpretation further in this article, relates to the treatment of effects and emotional states in the two hemispheres. In general, we know that the RH (Right Hemisphere) is more concerned with any kind of emotional arousal if compared with the LH (Left Hemisphere), which is specialized in other kinds of contents, such as planned actions or linguistic messages (for a general discussion, see Scardovelli 2017: Sec. 6.3). According to more subtle investigations, researchers could determine that the LH is concerned in particular with the spectrum of "positive" affects (pleasure, appreciation, desire), while "negative" affects (fear, depression, aversion) are more clearly associated with the RH (Powell and Schirillo 2009). In fact, researchers found that, in some art traditions, figures show a preferential orientation determined by the relationship—positive or negative—that the artists have with the subjects represented in the painting. Humphrey and McManus (1973), for example, analyzed 335 portraits created by Rembrandt. The persons portrayed have been divided into 5 categories: "female non-kin," "female kin," "male non-kin," "male kin," and "self-portraits." The authors noticed that if the subject represented was a stranger or a woman, it tended to face left. On the contrary, self-portraits tended to be right facing. In conclusion, the authors suggest that subjects facing left seem to reveal negative and aversive emotions, and subjects facing right seem to be linked with positive emotions.

Their conclusion is that the social distance from the artist determines an unconscious preference for left profile, associated with "social distance" and "negative effect," while the right profile is associated with "social proximity" and "positive effect." These authors do not associate this result with cerebral asymmetry. Nevertheless, other results based on laboratory studies openly associate better performances with left visual field and negative emotions, and right visual field with positive emotions (Jansari, Tranel, and Adolphs 2000; Powell and Schirillo 2009: 554–56; Reuter-Lorenz and Davidson 1981). But how does neural specialization (i.e., the hemispheric specialization for the treatment of effects), which has been studied in passive conditions, become responsible for an artistic creation? As I stated at the beginning of this section, following the insights of so-called "embodied cognition," it is possible to postulate analogous neurological mechanisms at work during passive and active situations. Thus, assuming that scenes and figures associated with negative effects are decoded and processed faster if located in the left visual field, it is possible to postulate that artists too will have a tendency to represent negative effects in the left hemifield and/or through a figure facing left (for a discussion, see Scardovelli 2017: Sec. 8.1). This is exactly what has been discovered in the study of Rembrandt's drawings. As anticipated, this factor will be the only one examined later in this article.

PROFILE PREFERENCE AND ANIMAL SPECIES

Statistical analyses permit us to go further in the study of orientation preferences in European parietal art. One of the most interesting studies that can be done concerns the variation of profile orientation among the different animal species represented on the cave walls (see Figs. 7-8).

Since a graphic is not sufficient for appreciating the statistical significance of the results, it is better to complete it with a crosstab, where the "Standardized Residual" allows us to appreciate how a certain frequency has to be considered significantly different from the expected result if data were distributed by chance. Standardized residuals that are greater than 2 mean that a specific result is significantly higher than the expected result, and those smaller than -2 mean that a result is significantly lower than the expected result. In Table 2 we can thus confirm that some animal species have a significant preferential orientation. Considering the animals showing a preferential left orientation, in the first stance felines show a statistically significant left orientation (standardized residual = 3.2). The second interesting category of data comes from the caprines, showing a particular preference for left orientation as well. In this case, I don't think that the outcome really deserves further attention, since this result seems to be uniquely related to the large number of figures of caprine heads that, as we have seen

Fig. 7.
In this graph we can see the difference in profile preference among the different animal species represented in parietal art.

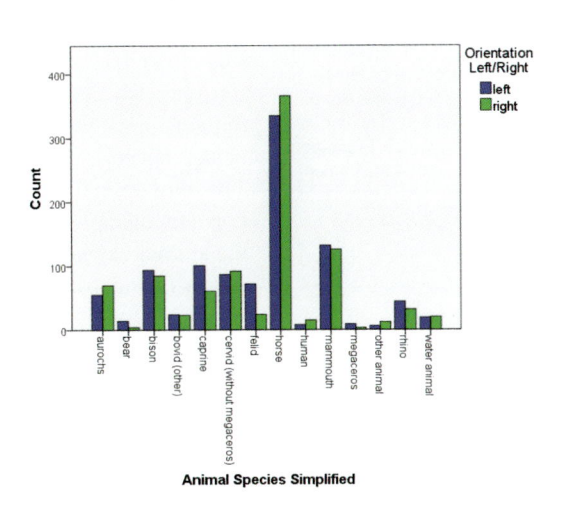

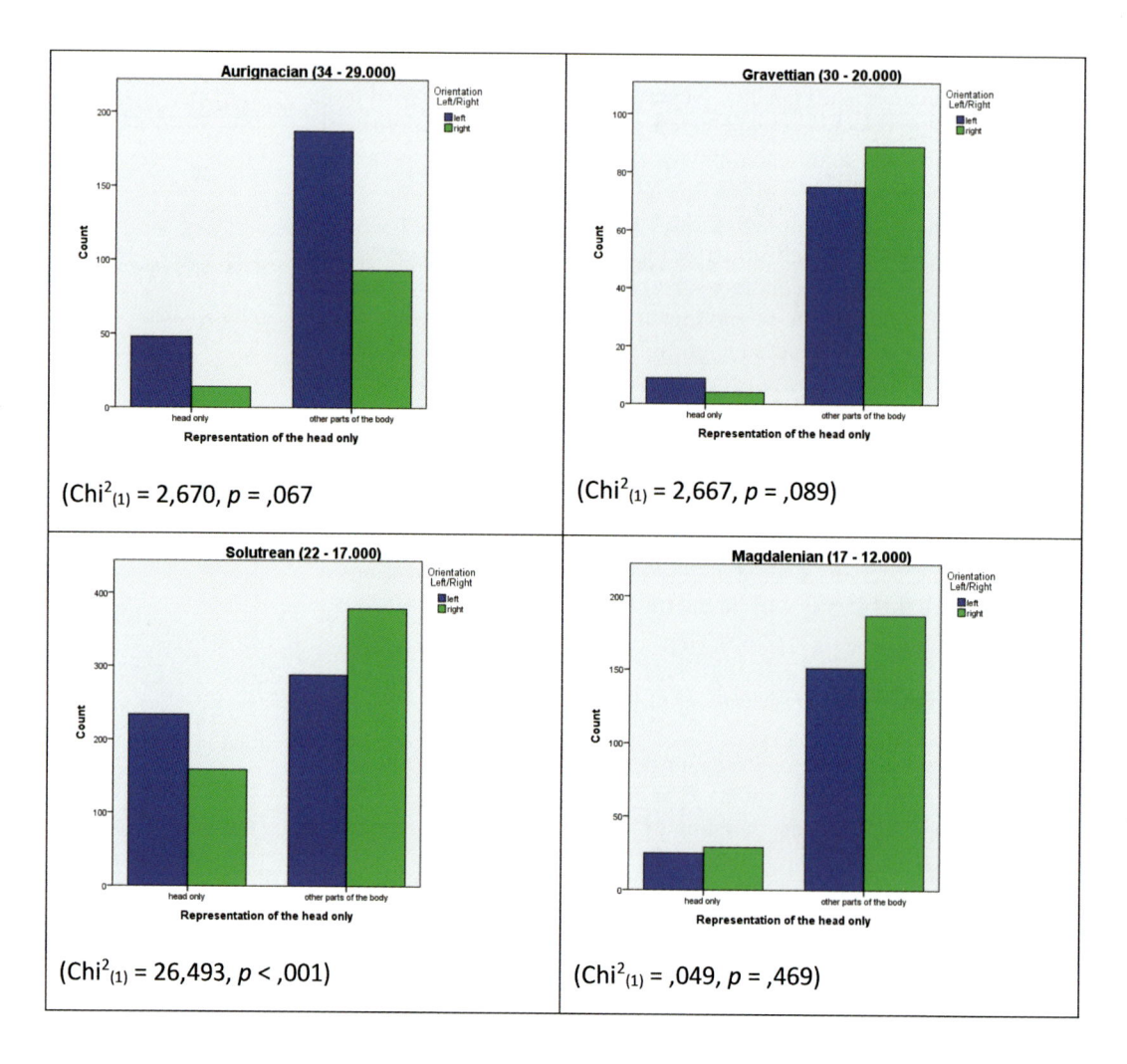

Fig. 8.
In this graphic it is possible to see how profile preference changes through time (here represented by Paleolithic cultures: Aurignacian, Gravettian, Solutrean, and Magdalenian) for four animal species, including bear, felid, horse and human.

Aurignacian (34 - 29.000): (Chi$^2_{(1)}$ = 2,670, p = ,067)

Gravettian (30 - 20.000): (Chi$^2_{(1)}$ = 2,667, p = ,089)

Solutrean (22 - 17.000): (Chi$^2_{(1)}$ = 26,493, p < ,001)

Magdalenian (17 - 12.000): (Chi$^2_{(1)}$ = ,049, p = ,469)

are preferentially represented in left profile due to mere neurological reasons. In fact, if we look at the right side of Table 2, where I considered all figures with the exception of heads only, we can see that the results remain almost the same for each animal category, with the only exception being caprines, which in this case do not have a preferential orientation. The third animal associated with preference for left profile is bear, which is oriented left 77% of the times (standardized residual = 1.5). In contrast, the animal categories showing a preference for right orientation include horses (standardized residual = 1.5) and humans (1.2).

Table 2.
In this table (left side) we can see how significantly the images of the different animal species are oriented in left or right profile (Chi$^2_{(13)}$ = 49,522, p < 0.001). On the right side of the table I consider all cases except the representations of heads only.

Animal Species Simplified		Orientation left/right crosstabulation			Orientation without animal heads		
		left	right	Total	left	right	Total
water animal	Count	19	20	39	19	20	39
	Percentage	48.7%	51.3%	100%	48.7%	51.3%	100%
	Standardized Residual	-0.3	0.3		0	0	
aurochs	Count	55	70	125	28	50	78
	Percentage	44%	56%	100%	35.9%	64.1%	100%
	Standardized Residual	-1.2	1.2		-1.6	1.6	
bison	Count	94	85	179	82	79	161
	Percentage	52.5%	47.5%	100%	50.9%	49.1%	100%
	Standardized Residual	0.1	-0.2		0.4	-0.4	
bovid (other)	Count	24	23	47	13	19	32
	Percentage	51.1%	48.9%	100%	40.6%	59.4%	100%
	Standardized Residual	-0.1	0.1		-0.6	0.6	
caprine	Count	101	61	162	55	51	106
	Percentage	62.3%	37.7%	100%	51.9%	48.1%	100%
	Standardized Residual	1.9	-1.9		0.5	-0.5	
cervid (without megaceros)	Count	87	93	180	61	71	132
	Percentage	48.3%	51.7%	100%	46.2%	53.8%	100%
	Standardized Residual	-0.6	0.7		-0.4	0.4	
horse	Count	335	366	701	188	249	437
	Percentage	47.8%	52.2%	100%	43%	57%	100%
	Standardized Residual	-1.5	1.5		-1.7	1.6	
felid	Count	72	24	96	43	22	65
	Percentage	75%	25%	100%	66.2%	33.8%	100%
	S Standardized Residual	3.2	-3.3		2.0	-2.0	
human	Count	8	15	23	6	14	20
	Percentage	34.8%	65.2%	100%	30%	70%	100%
	Standardized Residual	-1.1	1.2		-1.2	1.2	
mammoth	Count	132	126	258	128	114	242
	Percentage	51.2%	48.8%	100%	52.9%	47.1%	100%
	Standardized Residual	-0.1	0.1		1.0	-0.9	
megaceros	Count	9	3	12	9	3	12
	Percentage	75%	25%	100%	75%	25%	100%
	Standardized Residual	1.1	-1.2		1.3	-1.3	
bear	Count	14	4	18	12	3	15
	Percentage	77.8%	22.2%	100%	80%	20%	100%
	Standardized Residual	1.5	-1.6		1.7	-1.7	
rhino	Count	44	32	76	43	28	71
	Percentage	57.9%	42.1%	100%	60.6%	39.4%	100%
	Standardized Residual	0.7	-0.8		1.5	-1.4	
other animal	Count	7	12	19	5	11	16
	Percentage	36.8%	63.2%	100%	31.3%	68.8%	100%
	Standardized Residual	-0.9	0.9		-1.0	1.0	
Total	Count	1001	934	1935	692	734	1426
	Percentage	51.7%	48.3%	100%	48.5%	51.5%	100%

Aurignacian

Four animal species * Orientation Left/Right Crosstabulation

		Orientation Left/Right		Total
		left	right	
Bear	Count	9	3	12
	% within Four animal species	75,0%	25,0%	100,0%
	Standardized Residual	,1	-,1	
Felid	Count	51	18	69
	% within Four animal species	73,9%	26,1%	100,0%
	Standardized Residual	,0	-,1	
Horse	Count	39	15	54
	% within Four animal species	72,2%	27,8%	100,0%
	Standardized Residual	-,1	,2	
Human	Count	1	0	1
	% within Four animal species	100,0%	0,0%	100,0%
	Standardized Residual	,3	-,5	
Total	Count	100	36	136
	% within Four animal species	73,5%	26,5%	100,0%

(Chi$^2_{(3)}$ = .426, p = .935)

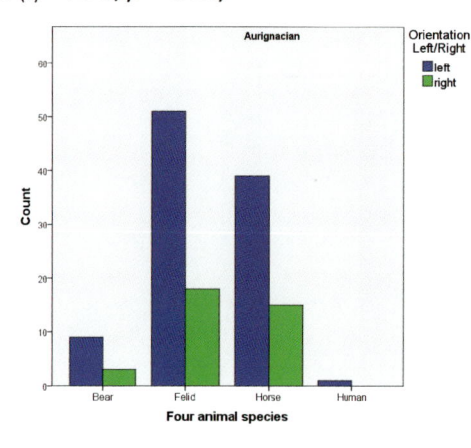

Fig. 9a.
Here we can see, for each of the chrono-culutral units (above, the Aurignacian) of the Upper Paleolithic, a crosstab showing the preference in profile of four animal categories (bear, felid, horse and human). Each crosstab is accompanied by a graph.

It is quite interesting to note that in the category of the left-oriented figures we find preferentially felines and bears, i.e., two predators (bears do not hunt all the time, being omnivores, but nonetheless they do hunt). Felines in particular represent a peculiar category in Upper Paleolithic visual imaginary, a topic that has been investigated since the beginning of the research on Paleolithic art (for reviews, see Fritz et al. 2011; Rousseau 1967). It is necessary to go back to what I said in the first section, when I quoted some studies suggesting that left profiles might be associated with some sort of negative effect, and right profiles with positive ones. In fact, it is possible that a preference for the left profile in felines might reflect a deep emotional connotation related with negative effects (for further details, see Scardovelli 2017: Chap. 8).

It is in fact astonishing that the only two terrestrial predators present in Paleolithic bestiary share the same preference for left profiles. On the other hand, we find horses and humans sharing a preference for right orientation. The bond connecting these two "animals" is less clear. Yet, ethologically speaking, they can both be considered as unique among Paleolithic bestiary. First of all, we can consider humans as "special," if not by definition, at least for the special treatment that human figures have received in Paleolithic art (Gaussen 1993). And if we have to find what distinguishes horses from the rest of Paleolithic bestiary, it is the only large herbivore present in Upper Paleolithic Europe without defensive means, such as horns, antlers, tusks, etc. Their specialization when facing danger is their velocity for running away. It is possible that this fact could have played a role in the construction of Paleolithic bestiary, revealing an interest in animal anatomy, horns, antlers, etc. But in conclusion, it is more reasonable to say that this second category, regarding humans and horses, is less clearly delineated than the first one. It is, however, interesting to consider that a similar dichotomy in orientation is found in the research of Le Quellec (1998) on the Neolithic

North African engravings of the Lybian Messak. The corpus of images reveals 10,000 units, and the figures on average are oriented to the right 71.9% of the time. In Le Quellec's table on page 322, equids are second (after ostriches) in right-facing species (79%), and the felids are second (after ovines) in left-facing species (63%), in relative and not absolute terms. This of course is not proof of anything, but still the parallelism deserves to be mentioned, also when we consider the large quantity of images studied by Le Quellec.

Figure 9a-c (divided into three parts across the following pages) shows how the preferential profile of these four animal species (bears, felids, humans and horses) changed over time. In fact, we already noticed (Fig. 2) that during the Aurignacian all animal species reveal a preference for the left profile (Fig. 3 showed that this trend is valid for Chauvet as well as for minor sites). The real shift occurred between the Aurignacian and the following periods, when we see the appearance of a sort of dichotomy. On the one hand, we find felids revealing a clear preference for the left profile (the statistical tests show a trend towards statistically significant results; data are in the table) followed by bears (whose numbers are too scarce for having any statistical significance). On the other hand, we find the other two animal categories (humans and horse) revealing a slight preference for the right profile, or no preference at all. Felids are the only category revealing a difference in statistical terms. In other words, during the Aurignacian, in all the caves that I recorded, all animal figures present a global preference for the left profile. From the Gravettian on, felids in particular maintain this preference, while the rest of the bestiary shift either to no real preference for a particular profile, with a slight preference for the right profile (horses and humans).

The information presented above provides the primary results of this study. We saw that in the passage from the Aurignacian to later periods a major shift occurs in some aesthetic value of images. What is astonishing is the sta-

Gravettian

Four animal species * Orientation Left/Right Crosstabulation

		Orientation Left/Right		Total
		left	right	
Felid	Count	3	0	3
	% within Four animal species	100,0%	0,0%	100,0%
	Standardized Residual	1,4	-1,3	
Horse	Count	19	20	39
	% within Four animal species	48,7%	51,3%	100,0%
	Standardized Residual	,2	-,2	
Human	Count	4	10	14
	% within Four animal species	28,6%	71,4%	100,0%
	Standardized Residual	-1,0	,9	
Total	Count	26	30	56
	% within Four animal species	46,4%	53,6%	100,0%

($Chi^2_{(2)}$ = 5,339, p = ,069)

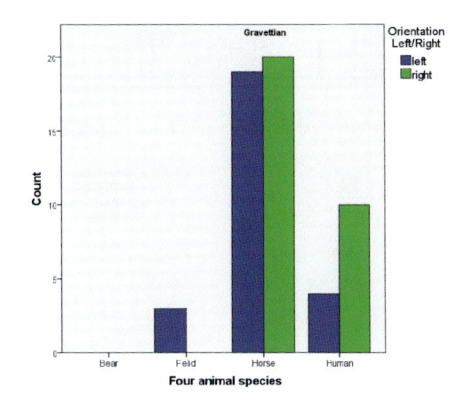

Fig. 9b.
The Gravettian.

Solutrean

Four animal species * Orientation Left/Right Crosstabulation

		Orientation Left/Right		Total
		left	right	
Bear	Count	3	1	4
	% within Four animal species	75,0%	25,0%	100,0%
	Standardized Residual	,8	-,8	
Felid	Count	15	6	21
	% within Four animal species	71,4%	28,6%	100,0%
	Standardized Residual	1,6	-1,5	
Horse	Count	254	297	551
	% within Four animal species	46,1%	53,9%	100,0%
	Standardized Residual	-,3	,3	
Human	Count	1	3	4
	% within Four animal species	25,0%	75,0%	100,0%
	Standardized Residual	-,6	,6	
Total	Count	273	307	580
	% within Four animal species	47,1%	52,9%	100,0%

($Chi^2_{(3)} = 7,245$, $p = ,064$)

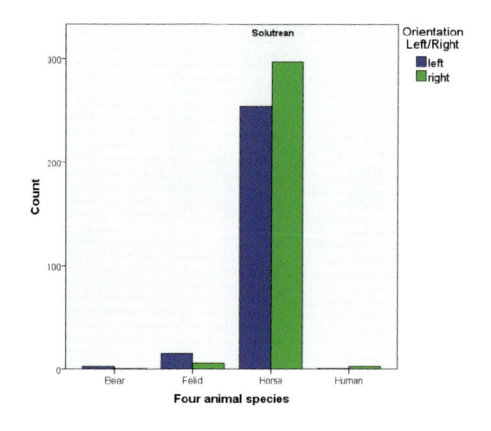

Fig. 9c.
The Solutrean.

bility of exactly the same trend beginning in the Gravettian, as if the Upper Paleolithic in Europe was divided, from the point of view of figure orientation and of artistic expression, in two major periods: the Aurignacian and all later periods. From a broader perspective, it is also interesting that Aurignacian artistic expression in general is characterized by a wide use of ivory-carving in realizing three-dimensional statuettes, while during the following periods, preferences in mobile art become the engravings on portable objects.

In conclusion, in this paper I examined the statistical distribution of left- and right-facing animal figures in the parietal art of the Upper Paleolithic. I noticed a major shift of profile preference between the Aurignacian and subsequent periods. This shift is so significant that, on a first level of interpretation, it seems to give credit to those models interpreting the passage from the Aurignacian to the Gravettian as a saltation scenario, as for example, in the context of a population replacement, and not as a slow cultural transformation (Otte 2011). More subtle analyses also seem to suggest that a strong dichotomy emerged starting with the Gravettian, a dichotomy that sees "dangerous" left-facing animals on one side (namely, felids and bears), and "non-dangerous" right-facing animals on the other (horses and humans). Further analyses are required to confirm or reject this hypothesis.

REFERENCES

Barrière, C. 1997. L'art pariétal des Grottes Des Combarelles. *Paléo* 1: 1–607.

Braswell, G. S., and K. K. Rosengren. 2000. Decreasing variability in the development of graphic production. *International Journal of Behavioral Development* 24 (2): 153–66.

Cashmore, L., N. T. Uomini, and A. Chapelain. 2008. The evolution of handedness in humans and great apes: A review and current issues. *Journal of Anthropological Sciences = Rivista Di Antropologia: JASS* 86: 7–35.

Chateau, J. 1965. *Attitudes intellectuelles et spatiales dans le dessin*. Paris: Centre national de la recherche scientifique.

Darras, B. 1996. *Au commencement était l'image: Du dessin de l'enfant à la communication de l'adulte*. Paris: ESF.

Fagard, J. 2004. *Droitiers/Gauchers: Des asymétries dans tous les sens*. Bruxelles: Groupe de Boeck.

Frayer, D. W., C. Lalueza-Fox, and L. Bondioli. 2010. Right handed Neandertals: Vindija and beyond. *Journal of Anthropological Sciences* 88: 113–27.

Frayer, D. W., M. Lozano, J. M. Bermúdez de Castro, E. Carbonell, J. L. Arsuaga, J. Radovčić, I. Fiore, and L. Bondioli. 2012. More than 500,000 years of right-handedness in Europe. *Laterality: Asymmetries of Body, Brain and Cognition* 17 (1): 51–69.

Fritz, C., P. Fosse, G. Tosello, G. Sauvet, and M. Azéma. 2011. Ours et lion: Réflexion sur la place des carnivores dans l'art Paléolithique. In *ResearchGate*, ed. by J.-P. Brugal, A. Gardeisen, and A. Zucker, pp. 299–318. Antibes: Éditions APDCA.

Gazzaniga, M. S., and C. S. Smylie. 1983. Facial recognition and brain asymmetries: Clues to underlying mechanisms. *Annals of Neurology* 13 (5): 536–40.

Groenen, M. 1997. La lateralizzazione nelle representazioni di mani negative Paleolitiche/La latéralisation dans les représentations des mains négatives Paléolithiques. *Manovre* 14: 31–59.

Guillaumin, J. 1961. Quelques faits et quelques réflexions à propos de l'orientation des profils humains dans les dessins d'enfants. *Enfance* 14 (1): 57–75.

Hinz, Berthold. 1974. Studien zur Geschichte des Ehepaarbildnisses. *Marburger Jahrbuch Für Kunstwissenschaft* 19: 139–218.

Humphrey, N. K., and I. C. McManus. 1973. Status and the left cheek. *New Scientist* 59: 437–39.

Jansari, A., D. Tranel, and R. Adolphs. 2000. A valence-specific lateral bias for discriminating emotional facial expressions in free field. *Cognition and Emotion* 3 (14): 341–53.

Le Quellec, J.-L. 1998. *Art rupestre et préhistoire du Sahara*. Paris: Payot.

McManus, I. C., and N. K. Humphrey. 1973. Turning the left cheek. *Nature* 243 (5405): 271–72.

Otte, M. 2011. La Gravettien, considéré en 2010. In *À la recherche des identités Gravettiennes: Actualités, questionnements et perspectives (Actes de la table ronde sur le Gravettien en France et dans les pays limitrophes. Aix-En-Provence, 6-8. X. 2008)*, by N. Goutas, L. Klaric, D. Pesesse, and P. Guillermain. Mémoire LII. Joué-Lès-Tours: Société Préhistorique Française.

Powell, W. R., and J. A. Schirillo. 2009. Asymmetrical facial expressions in portraits and hemispheric laterality: A literature review. *Laterality* 14 (6): 545–72. https://doi.org/10.1080/13576500802680336.

Reuter-Lorenz, P., and R. J. Davidson. 1981. Differential contributions of the two cerebral hemispheres to the perception of happy and sad faces. *Neuropsychologia* 19 (4): 609–13. https://doi.org/10.1016/0028-3932(81)90030-0.

Rousseau, M. 1967. *Les grands félins dans l'art de notre préhistoire*. Paris: Éditions A. & J. Picard et Cie.

Sauvet, G. 2005. La Latéralisation Des Figures Animales Dans Les Arts Rupestres: Un Exemple de Toposensitivité. Munibe. *Antropologia-Arkeologia* 57: 79–93.

Scardovelli, M. W. 2017. Étude sémiotique sur la latéralisation des figures animales dans l'art pariétal du Paléolithique En France. Montréal, Québec (Canada): Université du Québec à Montréal. https://www.academia.edu/36786618/%C3%A9tude_s%C3%A9miotique_sur_la_lat%C3%A9ralisation_des_figures_animales_dans_lart_pari%C3%A9tal_du_Pal%C3%A9olithique_en_France.pdf.

Simounet, C. 1975. La droite et la gauche dans le dessin de l'enfant et de l'adulte. *Enfance* 28 (1): 47–69.

Tosun, S., and J. Vaid. 2014. What affects facing direction in human facial profile drawing? A meta-analytic inquiry. *Perception Abstract* 43 (12): 1377–92.

Toth, N. 1985. Archaeological evidence for preferential right-handedness in the Lower and Middle Pleistocene, and its possible implications. *Journal of Human Evolution* 14 (6): 607–14.

Uhrbrock, R. S. 1973. Laterality in art. *The Journal of Aesthetics and Art Criticism* 32 (1): 27–35. https://doi.org/10.2307/428700.

Uomini, N. T. 2009. The prehistory of handedness: Archaeological data and comparative ethology. *Journal of Human Evolution* 57 (4): 411–19.

Van Sommers, P. 1984. *Drawing and cognition: Descriptive and experimental studies of graphic production processes*. New York: Cambridge University Press.

Zazzo, R. 1950. Le geste graphique et la structuration de l'esprit. *Enfance* 3 (1): 204–220.

Chapter 4

High level connections as a key component for the rapid dispersion of the Neolithic in Europe

Solange Rigaud[1]

Abstract

The transition to farming represents the process by which humans switched from hunting and gathering wild resources to a reliance on domesticated plants and animals. The adoption of domestication and sedentary life was probably promoted by a new system of beliefs and a profound reconfiguration of symbolic and social codes. This paper aims to present how personal ornaments inform the social reorganization of communities by tracking the multiple forms of interactions between groups and individuals. Technological and use-wear analysis of personal adornments, combined with the analysis of a georeferenced database of the bead types used by the last foragers and the first farmers in Europe, explores how interactions and communication networks led to the social reconfiguration of cultural groups and reshaped the cultural geography of Europe 8,000 years ago. The circulation of personal ornaments contributed to building and maintaining extensive and persistent networks of communication between hunter-gatherers and farmers. Long-term stability of contacts enabled the circulation of social, technical, and economic information, essential for the diffusion of the farming lifestyle. The long-term persistence of personal attires within farming communities suggests beads reflected the most entrenched and lasting facets of a farmer's identity compared to other cultural proxies.

INTRODUCTION

The Neolithic Revolution represents the process by which human groups switched from hunting and gathering wild resources to a reliance on systems of food production based on domesticated plants and animals. The reasons for this transformation, which occurred independently and at dif-

---

[1] CNRS, UMR 5199 – PACEA, Université de Bordeaux, Pessac, France.

© 2021, Kerns Verlag / https://doi.org/10.51315/9783935751377.004
Cite this article: Rigaud, S. 2021. High level connections as a key component for the rapid dispersion of the Neolithic in Europe. In *Ancient Connections in Eurasia*, ed. by H. Reyes-Centeno and K. Harvati, pp. 73-90. Tübingen: Kerns Verlag. ISBN: 978-3-935751-37-7.

ferent times in various regions of the world, have been debated for decades and are still not fully understood (Barker 2006; Bellwood 2005). Proposed causes include climate change (Gronenborn 2009; Richerson, Boyd, and Bettinger 2001; Rowley-Conwy and Layton 2011; Weninger et al. 2006), human–plant co-evolution (Rindos 1984), demography (Bocquet-Appel 2002; Bowles 2011), social incentives (Dietrich et al. 2012), competition and inequality (Wright 2014), or a combination of these (Ammerman and Biagi 2003; Bocquet-Appel 2008). Despite considerable debate concerning proposed causes and mechanisms, consensus exists that this revolution helped create the economic and social foundations on which present-day societies are based, such as diversified food production and storage techniques, surpluses, sedentism, labor specialization, social complexity, and ultimately state institutions.

In the Fertile Crescent, farming, herding, and sedentism progressively took place 12,000 years ago (12 ka), then spread across Europe from 8.8 ka until 5.5 ka (Ammerman and Cavalli-Sforza 1984; Bar-Yosef 2004; Pinhasi, Fort, and Ammerman 2005). Increasingly refined archaeological (Özdoğan 2011; Perlès 2003; Tresset and Vigne 2007), anthropological (Bocquet-Appel 2008; Fernández et al. 2014; Galeta et al. 2011; Lazaridis et al. 2014), and chronological data (Bocquet-Appel et al. 2009) identify a succession of profound cultural, technical, and economic changes between the last indigenous hunter-gatherers and the first Early Neolithic farmers in Europe. Recent genetic studies reveal complex demographic events took place during the three millennia that farming spread across Europe, including multiple inputs from farmers originating from the Near East, and also a contribution from local foragers and agriculturalist societies (Brandt et al. 2013; Galeta et al. 2011; Haak et al. 2010; Malmström et al. 2015). The transition to farming was not a linear process and it was slowed down, stopped, or even abandoned several times in specific regions before being definitively adopted in many areas (Shennan et al. 2013; Vigne et al. 2011). Along with these changes, it is generally recognized that the switch to agriculture resulted in, at least during the initial phases, more intense labor, a less diversified diet, increased morbidity, decreased life expectancy, precarious household-based production systems, and increased intra- and inter-group conflict (Cohen 2008; Hershkovitz and Gopher 2008).

Despite these potential disadvantages, after 5,000 years, the transition to farming was a success almost everywhere in Europe. Maintaining domestication, husbandry, and related cultural practices over large territories implies interindividual interactions with substantial transfers of knowledge (Larson et al. 2014). Beyond the skills required to select, reproduce, and raise animals and plants, domestication also signifies deep social and cultural changes within communities (Cauvin 1998; Digard 1988). Sedentariness may have also significantly affected contact to other people, the range of contacts, the way contacts were maintained between individuals and groups, and the dynamics of exchange of materials and ideas. It is likely that these profound economic and social changes transformed the way individuals perceived themselves and recognized each other and thus completely renewed their multiple past identities.

Identifying the material evidence of the multiple identities that may have existed within past foraging communities has puzzled archaeologists for several decades (Insoll 2007). Identities are linked to a broad cultural context, are socially mediated, and are implemented through embodiment, personal choices, and actions. They may refer both to individual identity and to group identity. The relation between individuals is seen as an essential factor conditioning people's actions, based on inexplicit rules and principles that guide their practices within society (Bourdieu 1977). Relationships between people are reproduced during a wide range of everyday activities, encompassing all aspects of the economical, technical, and ritual organizations of society. Group and personal identities are produced and maintained through the social processes related to these daily activities (Diaz-Andreu et al. 2005). The intrinsic link between an individual's activities and material productions, and interactions between people and identities, implies that the study of past material culture and related practices is insightful for the exploration of past identity constructions and changes (Dobres and Robb 2000). Approaches employed to explore the production and negotiation of identity in the archeological record rely on a broad range of cultural proxies, including tools, plant and animal exploitation, settlement organization, and body modification (Cobb and Gray Jones 2018; Finlay 2006). In this chapter, I will emphasize the recent conceptual, theoretical, and methodological developments in the exploration of past identities through the study of personal ornaments during the transition to the Neolithic in Europe.

Exactly like today, prehistoric personal ornaments transmitted symbolic messages in order to mediate the many social conventions related to individual and group identity (Sanders 2002). They were used for social transactions, rituals, the transmission of social memories, and to display social status within communities (Carter and Helmer 2015). Body ornaments were central to the creation of social and self identity. Their various associations and the way they were displayed on the body contributed to negotiating identities and unifying or distinguishing communities (Ogundiran 2002).

Numerous ethnographical studies have demonstrated that symbolic codes expressed by the association of ornaments on the human body change as a result of demic and cultural phenomena, including population replacement and admixture, trade, and the long distance diffusion of cultural traits (Lock and Symes 1999; Verswijver 1986). Personal ornaments can therefore be considered a reliable proxy for reconstructing cultural diversity and change in past societies. Here, I used personal ornaments to track possible interactions and contact networks during the Neolithic transition in Europe that led to changes in past cultural identities (Fig. 1).

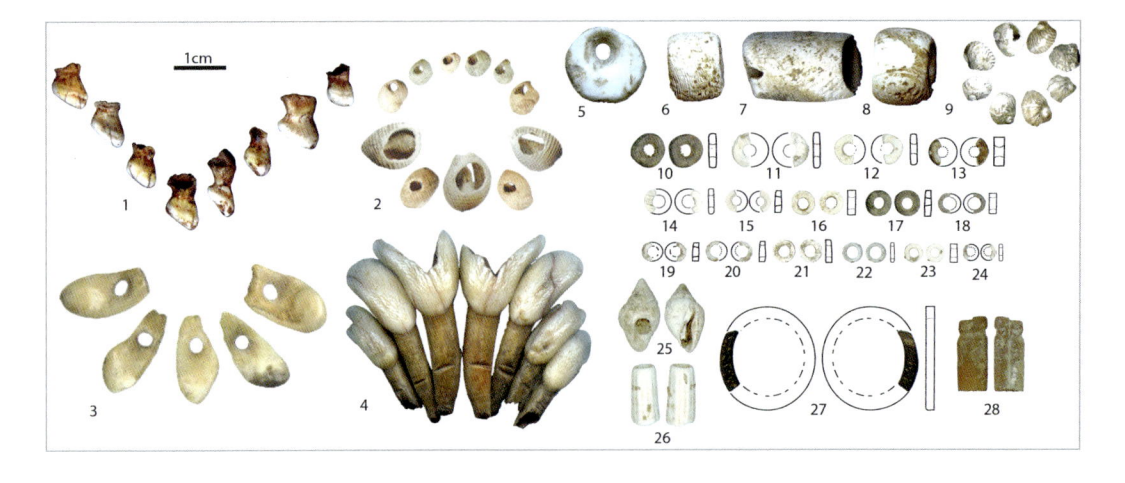

Fig. 1.
Selection of personal ornaments found in Mesolithic (1-4) and Early Neolithic contexts (5-28) showing a small part of the diversity of raw materials, shapes and techniques of suspensions potentially used to display various symbolic messages on the body. 1: Cyprinid teeth, Hohlenstein-Stadel (Germany), 2: *Littorina obtusata* and *Trivia* sp., El Mazo (Spain), 3: red deer canines, La Braña (Spain), 4: red deer incisors, Vedbaek-Bøgebakken (Denmark), 5-9: Essenbach-Ammerbreite (Germany), 5: calcite pendant, 6-8: *Spondylus* sp. beads, 9: *Theodoxus danubialis*, 10-28: Le Taï (France), 10: stone bead, 11-24: *Cerastoderma* sp. discoid beads, 25: *Columbella rustica*, 26: *Dentalium* sp., 27: stone ring, 28: calcite pendant (Modified after Rigaud 2014, 2013; Rigaud et al. 2018, 2013; Rigaud and Gutiérrez-Zugasti 2016; n°4 personal data see also Petersen et al. 2015).

HIGHLY CONNECTED MESOLITHIC FORAGING SOCIETIES

The transfer of cultural traits within and between past communities has long been investigated. Contact between populations can be seen through the circulation of raw materials from their source up to regions located several hundred kilometers away (Astruc 2011; Bajnóczi et al. 2013; Frost et al. 2004; Querré et al. 2014). The sharing of common stylistic traits in pottery design (Budja 2009; Hallgren 2004; Manen 2002), bone tools shaping and decoration (Man-Estier and Paillet 2013; Tartar et al. 2006), as well as flint weaponry production (Guilaine and Briois 2005; Langlais et al. 2016; Marchand 2003) are also commonly used to track interactions between communities. Maps describing connections, circulation roads, and exchange networks have been produced for many periods of prehistory across many regions (Álvarez Fernández 2008; Eriksen 2002; Rigaud 2013).

In their pioneering study, Newell et al. (1990) examined the chronological and spatial diversity of body ornaments produced by Epipaleolithic and Mesolithic hunter-gatherers from Western Europe, and used ethnographic data and a set of statistical analyses to map the geography of the social, ethnic, and linguistic groups (Newell et al. 1990). In this study, the main geographical corridors seem to be a key parameter for contact networks, the circulation of raw materials, and the shaping of Mesolithic cultural geography.

One of the most puzzling examples is the use of cyprinid teeth as body ornaments in the Late Mesolithic of the Upper and Lower Danube regions. The use of this raw material for ornaments emerges at the same time in these two remote regions (Newell et al. 1990; Rigaud 2011), and clearly corresponds to a cultural innovation since no previous use of these teeth has been attested during the Paleolithic. In both regions, the fish teeth were acquired from the Danube and suspended with a string attached using an adhesive compound (Cristiani, Farbstein, and Miracle

2014; Rigaud et al. 2013). Comparison of the material culture, economical organization, and mobility pattern between the two regions shows no other significant similarity (Bonsall 2008; Borić 2008; Jochim 1998; Orschiedt 1998). Body ornaments seem to have been a key element that culturally connected these two remote Danubian populations. Numerous other vast circulation roads lasted several millennia during the Mesolithic and at the beginning of the Neolithic. This is particularly the case for Mediterranean shells that circulated along the Rhône Valley up to Southern Germany (Álvarez-Fernández 2001) and along the Ebro Valley up to the Iberian Atlantic coast (Martinez-Moreno, Mora, and Casanova 2010) (Fig. 2).

However, contact networks were mostly not unidirectional. The use of modern reference data to study the 188 perforated red deer canines discovered in the multiple burials attributed to the final Mesolithic at Große Ofnet (Bavaria, Germany) shows an accumulation of the canines over time through a vast circulation network. Zooarchaeological and biogeographic data from modern and past reference samples suggests that the metric variability of the red deer canines accumulated at Große Ofnet covers the metric variability of red deer occupying Southern, Western, and Eastern Europe (Rigaud 2013). This result shows that perforated canines were accumulated through a multidirectional acquisition network.

Fig. 2.
Distribution of the Mediterranean *Columbella rustica* shells found in 118 Mesolithic occupations (modified after Rigaud 2011) showing that the shells circulated along the Rhône Valley up to Southern Germany (Álvarez-Fernández 2001) and along the Ebro Valley up to the Iberian Atlantic coast (Martinez-Moreno et al. 2010).

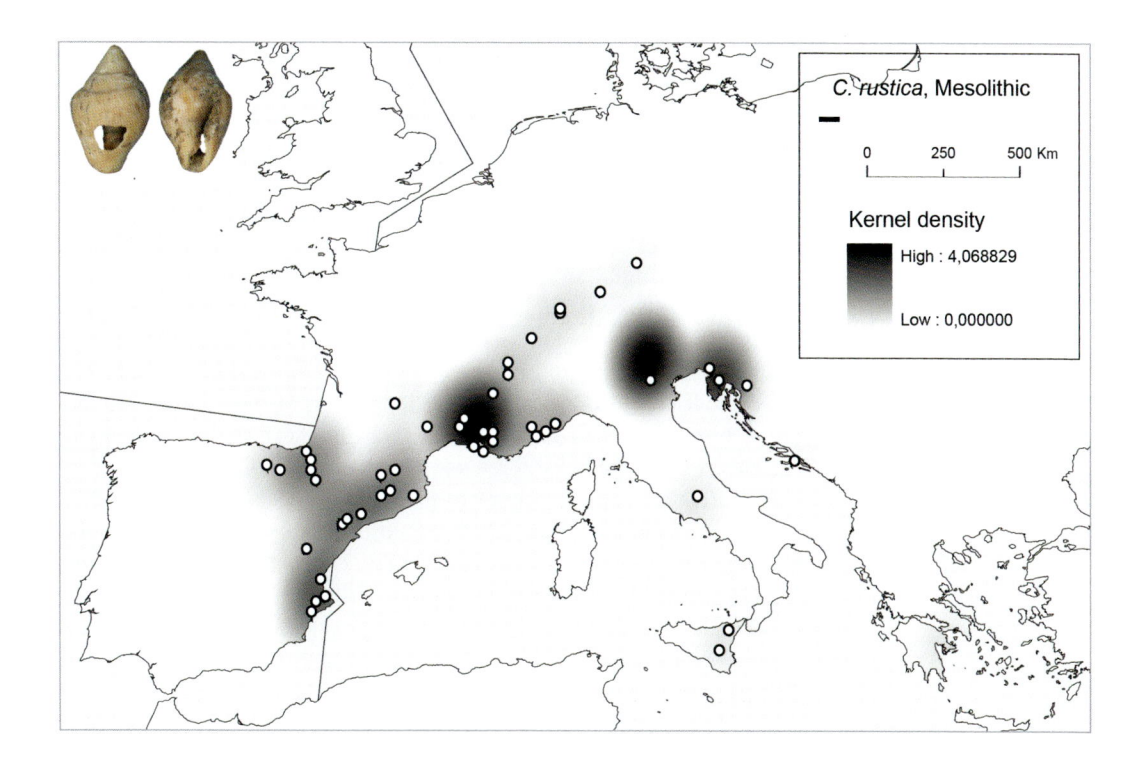

The cultural connections visible through the use of similar association of bead types also extends beyond cultural groups identified by other markers. In Atlantic Iberia, the identification of coastal and inland foraging societies who developed distinct economies (Arias 2005) and funerary rites (Arias Cabal 2007; Arias and Alvarez Fernandez 2004; Gutiérrez-Zugasti 2011) has led archaeologists to propose the existence of territoriality at the end of the Asturian Mesolithic. The presence of common associations of bead types within both societies, however, questions their cultural affinity. By investigating raw material procurement, selection strategies, and the manufacturing processes for shell bead production on coastal sites (El Mazo and El Toral, Spain), it has been proved that all the technical steps required for bead production were conducted in situ: collection of the shells, bead manufacture, and use (Rigaud and Gutiérrez-Zugasti 2016). Conversely, no evidence of shell bead manufacture was identified inland, suggesting the beads were premade before being introduced to the sites (Álvarez Fernández 2006; Arias and Álvarez Fernández 2004; Martınez 2004). Raw material sourcing combined with functional data highlights the complex interaction networks that existed during the Mesolithic between these two bounded communities, including the coastal communities, involved in shell bead production and spread, and the inland communities, which were geographically and economically disconnected from the coastal area but symbolically connected to the coastal group by their common personal ornaments (Rigaud and Gutiérrez-Zugasti 2016). This particular case study highlights the difficulty in bringing together economical, stylistic, and cultural data in order to define cultural groups.

DURABLE CONNECTIONS AND THE SHAPING OF THE EARLY NEOLITHIC CULTURAL GEOGRAPHY

This cultural substrate made up of ultra-connected Mesolithic communities may represent favorable conditions for enhancing the rapid dispersion of the Neolithic in Europe; however, beyond the material evidence of contacts between populations, mechanisms at work in the processes of cultural transmissions and diversifications have also been intensively studied. The contributions of Cavalli-Sforza and Feldman (1981) and Boyd and Richerson (1985), who applied models of evolutionary biology to the transmission of cultural traits, were pioneer studies (Boyd and Richerson 1985; Cavalli-Sforza and Feldman 1981). Cultural evolutionary theories rely on the statement that many aspects of interindividual and intergeneration transmissions are influenced by social learning and the cognitive capacities of human learners (learning and memory abilities) (Griffiths, Kalish, and Lewandowsky 2008). Cultural selection processes (best model, survival advantage), selection bias (efficiency, prestige and conformism in reproduction), and cultural drift (random choice) rule the emergence, persistence, and loss of cultural traits over time (Shennan 2011). Based on the model of "descent with modification" from ancestral populations, analysis of the pattern of variation in the archaeological record has contributed to the documentation of the various transmission mechanisms responsible for similarities and differences among groups in space and time (Collard and Shennan 2008; Jordan 2010; Shennan 2002; Shennan, Crema, and Kerig 2015; Tehrani and Collard 2009).

Within this analytical perspective, the database of Newell et al. (1990) was reassessed and updated in order to characterize the evolutionary mechanisms responsible for the spatial and chronological patterning of body ornaments during the Neolithic transition (Rigaud 2011). To conduct this study, archeological cultures were considered as the unit of analysis. Archaeological cultures correspond to geographic and chronological units characterized by archaeological occupations associated with durable material culture (Boyd and Richerson 1985; Lyman 2008) and represent a system of social information transmission that materializes population-level processes (Riede 2011). It is this short cut between archaeological cultures and past ethnicity that has led researchers to directly equate archeological cultures and past group identity (Childe 1962). This idea has since been widely debated (Binford 1965; Hodder 1978) by stressing that no consensus exists for the use of the concept of ethnicity to denote group versus individual and for the relation between ethnicity and its material expression (Banks 1996). Archeological cultures are mostly defined in the literature according to stone tool technology for the Paleolithic and Mesolithic periods and ceramic productions for the Neolithic.

This study relies on a database of the bead type associations identified in archaeological sites attributed to the three millennia during which the last hunter-gatherers and the first farmers interacted in Europe (Rigaud, d'Errico, and Vanhaeren 2015). It combines a series of multivariate analyses performed to characterize similarities and differences between archaeological cultures based on the diversity of bead type associations identified in each archaeological culture. Results indicate that the two main roads of Neolithic dispersal, through Central Europe and the Mediterranean, are characterized by distinct associations of bead types (Fig. 3). Personal ornaments from the Northern European regions are remarkably homogeneous compared to the highly diverse bead types present in Southern Europe.

Raw material availability, however, does not account for the observed pattern. The long distance trade of objects used as beads, well attested during the Mesolithic and Early Neolithic (Álvarez-Fernández 2001; Eriksen 2002; Martinez-Moreno, Mora, and Casanova 2010; Rigaud 2013; Zvelebil 2006), supplied the raw materials to regions where they were naturally rare or absent but where beads were still desired. The absence of amber ornaments outside the Baltic area cannot be attributed to the lack of this raw material: amber outcrops are documented in many regions of Europe (Czebreszuk 2007; Desailly 1930; Gardin 1986) and were exploited during the Upper Paleolithic (Beck, Chantre, and Sacchi

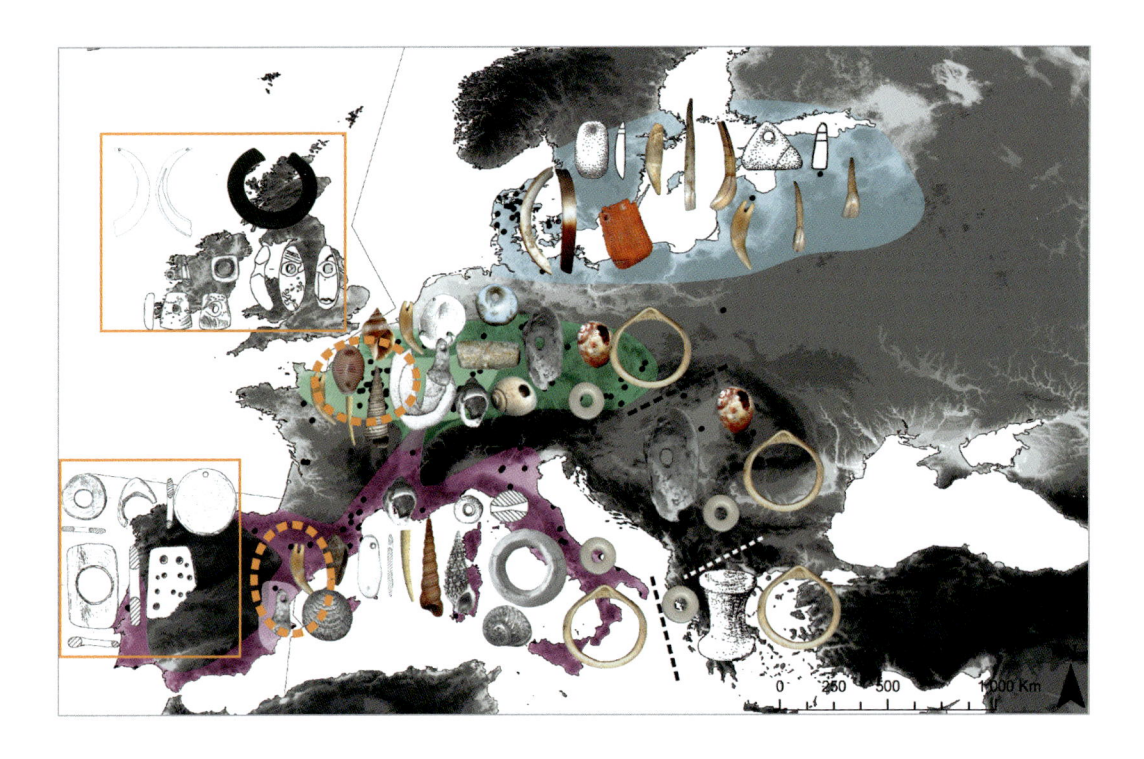

Fig. 3.
Cultural geography shaped by the Early Neolithic bead type associations identified in 488 archaeological occupations (black dots, modified after Rigaud 2011; Rigaud et al. 2015).

1987; White 2007) and probably the Bronze Age (Gardin 1986). Raw material availability also fails to explain the near complete absence of perforated shells in the Baltic area, where numerous suitable shell species were available, at least at the beginning of the transgression circa 8–7.2 ka, (Gutiérrez-Zugasti 2011; Høisæter 2009; Lewis 2011). Since raw material availability is not the determining factor for the observed pattern, the study concluded that the Mesolithic and Early Neolithic cultural geography identified by personal ornament diversity reflects cultural processes that drove the way individuals and groups identified themselves using bodily ornaments (Rigaud, d'Errico, and Vanhaeren 2015).

Bead type associations identify a well-defined and long-lasting stylistic boundary that persisted through time between Scandinavia and southernmost Europe (Rigaud, d'Errico, and Vanhaeren 2015). Population geneticists recently explored this frontier and identified two complete different population histories between Northern and Southern Europe. They concluded that specific migration patterns contributed to shape the Mesolithic material culture spatial patterning of Northern Europe (Jones et al. 2017; Malmström et al. 2009; Skoglund et al. 2014). The wide distribution of a specific personal ornament, namely perforated red deer canines, has also led geneticists to consider the high level of connection between groups as a major factor for the absence of genetic structure within southernmost European Mesolithic populations (Sánchez-Quinto et al., 2012). However, by considering an isolated bead type instead of the associations of personal ornaments, the authors have drastically neglected the high level of cultural diversity previously identified (Álvarez Fernández 2006; Dupont 2007; Newell et al. 1990; Rigaud 2011) and failed to relate population history to cultural geography.

The significant persistence of bead types used by hunter-gatherers is observed within farming communities in Central Europe and the Mediterranean region, where they are associated with new types of adornment exclusively present in farming communities. This is the case of the perforated *Columbella rustica* and other species of simply perforated gastropods in the Mediterranean area that are preserved from the Mesolithic to the Early Neolithic, but associated with new types of ornaments exclusively present within farming communities and previously unknown in Europe, in particular fully shaped objects such as discoid shell beads or stone bracelets. Bead type associations are highly diverse between each region during the beginning of the Neolithic, but show a similar trend in Central European and Mediterranean areas with the preservation of Mesolithic bead types combined with new Early Neolithic personal ornaments. These particular bead types, already identified in Mesolithic contexts, indicate that certain cultural traits, and probably also individuals, circulated from one society to another (Rigaud 2011; Rigaud, d'Errico, and Vanhaeren 2015). Genetic data (Bentley, Layton, and Tehrani 2009; Soares et al. 2010) are consistent with these results and identify complex demic processes, including the contribution of local hunter-gatherers and Near Eastern farmers to the European gene pool. The appropriation and incorporation of cultural traits could have facilitated the movement of individuals from one community to another and led to the persistence of cultural attributes initially adopted during the Mesolithic. This process could represent a successful strategy for farmers seeking to disperse in areas where large Mesolithic communities were already present, and implies that the cultural geography identified by personal ornaments at the beginning of the Neolithic in Europe is rooted in symbolic practices and stylistic choices inherited from the Mesolithic foraging communities, probably reflecting the most entrenched and lasting facets of a farmer's cultural identity (Rigaud, Manen, and García-Martínez de Lagrán 2018).

These data show that bead production at the dawn of the Neolithic reflects a long-term strategy favoring the replication of symbolic messages transmitted by personal ornaments. In addition to maintaining supply networks over time, a hypothesis formulated to explain the faithful reproduction of bead type associations suggests that specific transmission processes acted and involved a small number of specialized craftsmen responsible for bead manufacture within the first farming communities (Rigaud, Manen, and García-Martínez de Lagrán 2018). Involving few specialized crafters for bead production may limit errors in replication and secure long-term maintenance of styles and symbolic codes. This hypothesis opens up the possible existence of particular sites dedicated to the production and dispersal of ornaments holding a key place in the symbolic landscape of the communities. This category of site is rarely identified in the archaeological record but they are known from the end of the Paleolithic (Rigaud et al. 2019; 2014). At Franchthi Cave (Greece), a large amount of shells, ranging from unperforated in perfect condition to heavily used or broken ornaments, have been recovered in every Mesolithic unit (Perlès 2018). The presence of a large number of shells with use-wear suggests worn elements were removed and replaced by newly manufactured ones. Freshly embroidered garments were probably exported from the site (worn by the Franchthi inhabitants themselves) or exchanged with inland sites where similar bead types are found (Perlès 2018). This pattern defines Franchthi Cave as a lasting production center that drove the way people shaped their body ornamentation.

CONCLUSION

By studying the personal ornaments belonging to the last foraging and the first farming communities, I examined the circulation, exchange, and transmission of objects, as well as the aesthetic standards and symbols between groups relying on drastically different economies. More specifically, I explored how societies established symbolic relationship through the use of common associations of ornaments and the mechanisms that led some societies to adopt body ornamentations different from those of neighboring communities. The results highlight the stylistic, territorial, and symbolic identities of past human populations who occupied Europe at this time. Europe appears as a cultural patchwork where early farming communities faced different challenges, implying dissimilar opportunities for the exchange and transfer of information with foraging communities and for access to new territories. Changes in personal ornamentation show that population dynamics were ruled by the renewal of symbolic standards, linked to social norms and systems of belief.

Besides the clear impact of cultural transfers between populations, the role of environmental factors in shaping the cultural geography of Europe has also been characterized (Banks et al. 2013). By applying two predictive architectures to reconstruct the eco-cultural niches of farming populations, based on their geographic occurrences and abiotic and climatic data, ecological niche modeling indicates that cultural processes behind the spread of farming in Europe took place in specific environments compatible with particular cultural adaptations (Banks et al. 2013). That these processes of economic specialization took place in particular environments reinforces the idea of major adaptations of farming cultures within distinct environmental envelopes. Foraging societies were probably not passive participants in the European ecosystem (Colehour 2008): knowledge of seasonal fluctuations in the local environment and landscape, soil properties, patterns of natural germination of local wild plants, and water availability are all essential in the development of a successful production system. Social interactions between foraging and farming populations highlighted by the personal ornaments analysis may have often granted the transmission of useful naturalistic knowledge and related know-how which remain to be explored. Maintaining and reinforcing connections with neighboring communities represented an efficient strategy for emerging farming societies seeking to spread and access new territory.

ACKNOWLEDGMENTS

The author wishes to thank Katerina Harvati, Gerhard Jäger and Hugo Reyes-Centeno for the invitation to the symposium at the University of Tübingen. This project was successively funded by the Fyssen Foundation, The Marie Curie COFUND programme, the French National Research Agency's project ANR-13-CULT-0001-01 and the CNRS momentum project "Symbolling and Neighboring at the Dawn of Agriculture in Europe 8000 years ago." The author is also supported by the "Grand Programme de Recherche, Human Past" from the University of Bordeaux.

REFERENCES

Álvarez Fernández, E. 2008. The use of *Columbella rustica* (Clase: Gastropoda) in the Iberian Peninsula and Europe during the Mesolithic and Early Neolithic. In *V Congreso del Neolítico Peninsular (Alicante, 27-30 noviembre 2006)*, ed. by M. Hernández Pérez, J. A. Soler García, and J. A. López Padilla, pp. 103–11. Alicante: Museo Arqueológico de Alicante, Diputación Provincial de Alicante.

Álvarez Fernández, E. 2006. *Los objetos de adorno-colgantes del Paleolitico superior y del Mesolitico en la Cornisa Cantabrica y en el Valle del Ebro: una vision europea.* Salamanca: Universidad de Salamanca.

Álvarez-Fernández, E. 2001. L'axe Rhin-Rhône au Paléolithique supérieur récent: l'exemple des mollusques utilisés comme objets de parure. *L'Anthropologie* 105: 547–64.

Ammerman, A. J., and L. L. Cavalli-Sforza. 1984. *The Neolithic transition and the genetics of population in Europe.* Princeton: Princeton University Press.

Ammerman, A. J., and P. Biagi. 2003. *The Widening Harvest. The Neolithic Transition in Europe: Looking Back, Looking Forward.* Archaeological Institute of America. Boston.

Arias Cabal, P. 2007. Neighbours but diverse: social change in north-west Iberia during the transition trom the Mesolithic to the Neolithic (5500-4000 cal BC). *Proceedings of the British Academy* 144: 53–71.

Arias, P. 2005. Determinaciones de isótopos estables en restos humanos de la región cantábrica. Aportación al estudio de la dieta de las poblaciones del Mesolítico y el Neolítico. *MUNIBE (Antropologia-Arkeologia)* 57: 359–74.

Arias, P., and E. Alvarez Fernandez. 2004. Iberian Foragers and Funerary Ritual. A Review of Paleolithic and Mesolithic Evidence on the Peninsula. In *The Mesolithic of the Atlantic Façade: Proceedings of the Santander Symposium. Anthropological Research Papers no 55*, ed. by R. Gonzalez Morales and G. A. Clark, pp. 225–48. Tempe: Arizona State University.

Astruc, L. 2011. Du Göllüdağ à Shillourokambos: de l'utilisation d'obsidiennes anatoliennes en contexte insulaire. In *Shillourokambos, un établissement néolithique pré-céramique* à Chypre. Les fouilles du secteur 1, ed. by J. Guilaine, F. Briois, and J.-D. Vigne, pp. 727–44. Efa.

Bajnóczi, B., G. Schöll-Barna, N. Kalicz, Z. Siklósi, G. H. Hourmouziadis, F. Ifantidis, A. Kyparissi-Apostolika, M. Pappa, R. Veropoulidou, and C. Ziota. 2013. Tracing the source of Late Neolithic Spondylus shell ornaments by stable isotope geochemistry and cathodoluminescence microscopy. *Journal of Archaeological Science* 40 (2): 874–82. https://doi.org/10.1016/j.jas.2012.09.022.

Banks, M. 1996. Ethnicity: Anthropological Constructions. London: Routledge.

Banks, William, N. Antunes, F. D'Errico, and S. Rigaud. 2013. Ecological Constraints on the First Prehistoric Farmers in Europe. *Journal of Archaeological Science* 40 (6): 2746–2753.

Barker, G. 2006. *The Agricultural Revolution in Prehistory: Why did Foragers become Farmers?* Oxford University Press.

Bar-Yosef, O. 2004. Guest editorial: East to West - Agricultural origins and dispersal into Europe. *Current Anthropology* 45 (S4): S1–3.

Beck, C. W., F. Chantre, and D. Sacchi. 1987. L'"ambre" paléolithique de la grotte d'Aurensan (Hautes-Pyrénées). *L'Anthropologie* 91 (1): 259–61.

Bellwood, P. 2005. *First Farmers: The Origins of Agricultural Societies*. Blackwell, Oxford.

Bentley, R.A., R. Layton, and J. Tehrani. 2009. Kinship, Marriage, and the Genetics of Past Human Dispersals. *Human Biology* 81 (2-3): Article 4.

Binford, L. R. 1965. Archaeological systematics and the study of culture process. *Antiquity* 31: 203–10.

Bocquet-Appel, J.-P. 2002. Paleoanthropological Traces of a Neolithic Demographic Transition. *Current Anthropology* 43 (4): 637–50.

Bocquet-Appel, J.-P., S. Naji, M. V. Linden, and J. K. Kozlowski. 2009. Detection of diffusion and contact zones of early farming in Europe from the space-time distribution of 14C dates. *Journal of Archaeological Science* 36: 807–20.

Bocquet-Appel, J. P. 2008. The Neolithic demographic transition, population, pressure and cultural change. *Comparative Civilizations Review* 58: 36–49.

Bonsall, C. 2008. The Mesolithic of the Iron Gates. In *Mesolithic Europe*, ed. by G. Bailey and P. Spikins, pp. 238–79. Cambridge: Cambridge University Press.

Borić, Dušan. 2008. Lepenski Vir culture in the light of new research. *Journal of the Serbian Archaeological Society* 44: 9–44.

Bourdieu, P. 1977. *Outline of a Theory of Practice*. Cambridge Studies in Social and Cultural Anthropology. Cambridge: Cambridge University Press. https://doi.org/10.1017/CBO9780511812507.

Bowles, S. 2011. Cultivation of cereals by the first farmers was not more productive than foraging. *Proceedings of the National Academy of Sciences*. http://www.pnas.org/content/early/2011/03/02/1010733108.abstract.

Boyd, R., and P. Richerson. 1985. *Culture and the Evolutionary Process*. University of Chicago Press.

Brandt, G., W. Haak, C. J. Adler, C. Roth, A. Szécsényi-Nagy, S. Karimnia, S. Möller-Rieker, et al. 2013. Ancient DNA Reveals Key Stages in the Formation of Central European Mitochondrial Genetic Diversity. *Science* 342 (6155): 257–61.

Budja, M. 2009. Early Neolithic pottery dispersals and demic diffusion in Southeastern Europe. *Documenta Praehistorica* XXXVI: 117–37.

Carter, B., and M. Helmer. 2015. Elite Dress and Regional Identity: Chimú-Inka Perforated Ornaments from Samanco, Nepeña Valley, Coastal Peru. *BEADS: Journal of the Society of Bead Researchers* 20: 46–74.

Cauvin, J. 1998. *Naissance Des Divinités, Naissance De L'agriculture: La Révolution Des Symboles Au Néolithique*. Paris: Paris: CNRS.

Cavalli-Sforza, L. L., and M. W. Feldman. 1981. *Cultural Transmission and Evolution: A Quantitative Approach*. Princeton: Princeton University Press.

Childe, V. G. 1962. *L'Europe préhistorique: les premières sociétés européennes*. Paris: Payot.

Cobb, H., and A. Gray Jones. 2018. Being Mesolithic in Life and Death. *Journal of World Prehistory* 31 (3): 367–83. https://doi.org/10.1007/s10963-018-9123-1.

Cohen, M. N. 2008. Implications of the NDT for World Wide Health and Mortality in Prehistory. In *The Neolithic Demographic Transition and its Consequences*, ed. by J-P. Bocquet-Appel and O. Bar-Yosef, pp. 481–500. Springer.

Colehour, A. M. 2008. The Biogeography of Plant Domestication. *Macalester Reviews in Biogeography* 1 (Article 1): 1–26.

Collard, M., and S. J. Shennan. 2008. Patterns, processes and parsimony: Studying cultural evolution with analytical techniques from evolutionary biology. In *Cultural transmission and material culture: breaking down boundaries*, ed. by M. Stark, B. J. Bowser, and L. Horne, pp. 17–33. Tucson: University of Arizona Press.

Cristiani, E., R. Farbstein, and P. Miracle. 2014. Ornamental traditions in the Eastern Adriatic: The Upper Palaeolithic and Mesolithic personal adornments from Vela Spila (Croatia). *Journal of Anthropological Archaeology* 36 (0): 21–31. https://doi.org/10.1016/j.jaa.2014.06.009.

Czebreszuk, J. 2007. Amber between the Baltic and the Aegean in the third and second Millenia BC (an Outline of Major Issues). In *Between the Aegean and Baltic Seas. Proceedings of the International Conference Bronze and Early Iron Age Interconnections and Contemporary Developments between the Aegean and the Regions of the Balkan Peninsula, Central and Northern Europe*, ed. by I. Galanaki, H. Tomas, Y. Galanakis, and R. Laffineur: 363–68. Aegaeum 27. Zagreb.

Desailly, L. 1930. L'Ambre jaune fossile en France et en Belgique. *Bulletin de la Société préhistorique de France* 27: 360–62.

Diaz-Andreu, M., S. Lucy, S. Babic, and D. Edwards. 2005. *The archaeology of identity: Approaches to gender, age, ethnicity, status and religion*. London: Routledge.

Dietrich, O., M. Heun, J. Notroff, K. Schmidt, and M. Zarnkow. 2012. The role of cult and feasting in the emergence of Neolithic communities. New evidence from Göbekli Tepe, south-eastern Turkey. *Antiquity* 86 (333): 674–95. https://doi.org/10.1017/S0003598X00047840.

Digard, J.-P. 1988. Jalons pour une anthropologie de la domestication animale. *L'Homme* 28 (4): 27–58.

Dobres, M.-A., and J. E. Robb. 2000. *Agency in Archaeology*. London: Routledge.

Dupont, C. 2007. Les amas coquilliers mésolithiques de Téviec et d'Hoedic et le dépôt coquillier néolithique d'er Yoc'h : de la ressource alimentaire à l'utilisation des coquilles vides. *Melvan, La Revue des deux îles* 4: 251–64.

Eriksen, B. V. 2002. Fossil Mollusks and Exotic Raw Materials in Late Glacial and Early Find Contexts: A Complement to Lithic Studies. In *Lithic raw material economy in late glacial and early posglacial western Europe*, ed. by L. E. Fisher and B. Valentin Eriksen, pp. 27–52. Oxford: Bar International Series.

Fano Martınez, M. A. 2004. Un nuuevi tiempo: el Mesolítico en la región Cantábrica. *Kobie* 8: 337–402.

Fernández, E., A. Pérez-Pérez, C. Gamba, E. Prats, P. Cuesta, J. Anfruns, M. Molist, E. Arroyo-Pardo, and D. Turbón. 2014. Ancient DNA Analysis of 8000 B.C. Near Eastern Farmers Supports an Early Neolithic Pioneer Maritime Colonization of Mainland Europe through Cyprus and the Aegean Islands. *PLoS Genet* 10 (6): e1004401. https://doi.org/10.1371/journal.pgen.1004401.

Finlay, N. 2006. Mesolithic Britain and Ireland: New approaches. In *Mesolithic Britain and Ireland: New approaches*, ed. by C. Conneller and G. Warren, pp. 25–60. Tempus: Stroud.

Frost, R. L., M. L. Weier, K. L. Erickson, O. Carmody, and S. J. Mills. 2004. Raman spectroscopy of phosphates of the variscite mineral group. *Journal of Raman Spectroscopy* 35 (12): 1047–55. https://doi.org/10.1002/jrs.1251.

Galeta, P., V. Sládek, D. Sosna, and J. Bruzek. 2011. Modeling Neolithic dispersal in Central Europe: Demographic implications. *American Journal of Physical Anthropology* 146 (1): 104–15. https://doi.org/10.1002/ajpa.21572.

Gardin, C.du. 1986. La parure d'ambre à l'âge du Bronze en France. *Bulletin de la Société préhistorique française* 83: 546–88.

Griffiths, T. L., M. L. Kalish, and S. Lewandowsky. 2008. Theoretical and empirical evidence for the impact of inductive biases on cultural evolution. *Philosophical Transactions of the Royal Society B: Biological Sciences* 363 (1509): 3503. https://doi.org/10.1098/rstb.2008.0146.

Gronenborn, D. 2009. Climate fluctuations and trajectories to complexity in the Neolithic: Towards a theory. *Documenta Praehistorica* 36: 97–110.

Guilaine, J., and B. François. 2005. Shillourokambos et la néolithisation de Chypre: quelques reflexions. *Mayurqa* 30: 13–32.

Gutiérrez-Zugasti, I. 2011. Coastal resource intensification across the Pleistocene-Holocene transition in Northern Spain: Evidence from shell size and age distributions of marine gastropods. *Quaternary International* 244: 54–66.

Haak, W., Oleg Balanovsky, Juan J. Sanchez, Sergey Koshel, Valery Zaporozhchenko, ChristinaJ. Adler, Clio S.I Der Sarkissian, et al. 2010. Ancient DNA from European Early Neolithic Farmers RevealsTheir Near Eastern Affinities. *PLoS Biology* 8 (11): e1000536.doi:10.1371/journal.pbio.1000536.

Hallgren, F. 2004. The introduction of ceramic technology around the Baltic Sea in the 6 th millennium. In *Coast to coast- landing*, ed. by H. Knutsson, pp. 123–42. Uppsala.

Hershkovitz, I., and A. Gopher. 2008. Demographic, Biological and Cultural Aspects of the Neolithic Revolution: A View from the Southern Levant. In *The Neolithic Demographic Transition and its Consequences*, ed. by J-P. Bocquet-Appel and O. Bar-Yosef, pp. 441–80. Springer.

Hodder, I. 1978. Simple correlations between material culture and society: A review. In *The Spatial Organisation of Culture*, ed. by Ian Hodder, pp. 3–24. London: Duckworth.

Høisæter, T. 2009. Distribution of marine, benthic, shell bearing gastropods along the Norwegian coa. *Fauna norvegica* 28: 5–106.

Insoll, T. 2007. *The Archaeology of Identities: A Reader*. London: Routledge.

Jochim, M. A. 1998. *A Hunter-Gatherer Landscape: Southwest Germany in the Late Paleolithic and Mesolithic*. Springer. Interdisciplinary Contributions to Archaeology. New York: Springer.

Jones, E. R., G. Zarina, V. Moiseyev, E. Lightfoot, P. R. Nigst, A. Manica, R. Pinhasi, and D. G. Bradley. 2017. The Neolithic Transition in the Baltic Was Not Driven by Admixture with Early European Farmers. *Current Biology* 27 (4): 576–82. https://doi.org/10.1016/j.cub.2016.12.060.

Jordan, P. 2010. Understanding the spread of innovations in prehistoric social networks: new insights into the origins and dispersal of early pottery in Northern Eurasia. In *Transference: interdisciplinary communications*, ed. by W. Ostreing. Centre for Advanced Studies, internet publication at https://www.cas.uio.no/publications/transference.php. Oslo.

Langlais, M., A. Sécher, S. Caux, V. Delvigne, L. Gourc, C. Normand, and M. Sánchez de la Torre. 2016. Lithic tool kits: A Metronome of the evolution of the Magdalenian in southwest France (19,000–14,000 cal BP). *Quaternary International* 414: 92–107. https://doi.org/10.1016/j.quaint.2015.09.069.

Larson, G., D. R. Piperno, R. G. Allaby, M. D. Purugganan, L. Andersson, M. Arroyo-Kalin, L. Barton, et al. 2014. Current perspectives and the future of domestication studies. *Proceedings of the National Academy of Sciences USA* 111 (17): 6139–46. https://doi.org/10.1073/pnas.1323964111.

Lazaridis, I., N. Patterson, A. Mittnik, G. Renaud, S. Mallick, K. Kirsanow, P. H. Sudmant, et al. 2014. Ancient human genomes suggest three ancestral populations for present-day Europeans. *Nature* 513 (7518): 409–13.

Lewis, J. P. 2011. *Holocene environmental change in coastal Denmark: Interactions between land, sea and society*. Ph.D. Denmark: Loughborough University.

Lock, A., and K. Symes. 1999. Social relations, communication, and cognition. In *Human Symbolic Evolution*, ed. by A. Lock and C. R. Peters, pp. 204–32. Oxford: Oxford Science Publication.

Lyman, R. L. 2008. Cultural Transmission in North American Anthropology and Archaeology, ca. 1895-1965. In *Cultural Transmission and Archaeology: Issues and Case Studies*, ed. by M. J. O'Brien, pp. 10–20. Washington D.C.: Society for American Archaeology Press.

Malmström, H., M. T. P. Gilbert, M. G. Thomas, M. Brandström, J. Storå, P. Molnar, P. K. Andersen, et al. 2009. Ancient DNA Reveals Lack of Continuity between Neolithic Hunter-Gatherers and Contemporary Scandinavians. *Current Biology* 19: 1758–62.

Malmström, H., A. Linderholm, P. Skoglund, J. Storå, P. Sjödin, M. Thomas P. Gilbert, G. Holmlund, et al. 2015. Ancient mitochondrial DNA from the northern fringe of the Neolithic farming expansion in Europe sheds light on the dispersion process. *Philosophical Transactions of the Royal Society B: Biological Sciences* 370 (1660). https://doi.org/10.1098/rstb.2013.0373.

Manen, C. 2002. Structure et identité des styles céramiques du Néolithique ancien entre Rhône et Èbre. *Gallia préhistoire* 44: 121–65.

Man-Estier, E., and P. Paillet. 2013. Rochereill et l'art Magdalénien de la fin du Tardiglaciaire dans le nord du Périgord (Dordogne, France). In *Expressions esthétiques et comportements techniques au Paléolithique*, ed. by Marc Groenen, pp. 7–36. Oxford: Archeopress. BAR International Series.

Marchand, G. 2003. Les zones de contact Mésolithique / Néolithique dans l'ouest de la France : définition et implications. In *Muita gente, poucas antas? Origens, espaços e contextos do Megalitismo. Actas do II Coloquio Internacional sobre Megalitismo*, ed. by V. S. Gonçalves, pp. 181–97. Trabalhos de Arqueologia.

Martinez-Moreno, J., R. Mora, and J. Casanova. 2010. Lost in the mountains? Marine ornaments in the Mesolithic of the northeast of the Iberian Peninsula. *MUNIBE* 31: 100–109.

Newell, R. R., D. Kielman, T. S. Constandse-Westermann, W. A. B. van der Sanden, and A. Van Gijn. 1990. An Inquiry Into the Ethnic Resolution of Mesolithic Regional Groups: The Study of Their Decorative Ornaments in Time and Space. Leyden: Brill.

Ogundiran, A. 2002. Of Small Things Remembered: Beads, Cowries, and Cultural Translations of the Atlantic Experience in Yorubaland. *The International Journal of African Historical Studies* 35 (2/3): 427–57. https://doi.org/10.2307/3097620.

Orschiedt, J. 1998. Ergebnisse einer neuen Untersuchung der spätmesolithischen Kopfbestattungen aus Süddeutschland. In *Aktuelle Forschungen zum Mesolithikum*, ed. by N. Conard and J. C. Kind, pp. 147–60. Tubingen: Mo Vince.

Özdoğan, M. 2011. Archaeological Evidence on the Westward Expansion of Farming Communities from Eastern Anatolia to the Aegean and the Balkans. *Current Anthropology* 52 (S4): S415–30. https://doi.org/10.1086/658895.

Perlès, C. 2003. An alternate (and old-fashioned) view of Neolithisation in Greece. *Documenta Praehistorica* XXX: 99–113.

Perlès, C. 2018. *Ornaments and other ambiguous artifacts from Franchthi. Volume 1, The Palaeolithic and the Mesolithic. Excavations at Franchthi Cave, Greece*. Bloomington: Indiana University Press.

Pinhasi, R., J. Fort, and A. J. Ammerman. 2005. Tracing the Origin and Spread of agriculture in Europe. *PLoS Biology* 3 (12): 2220–28.

Querré, G., T. Calligaro, S. Domínguez-Bella, and S. Cassen. 2014. PIXE analyses over a long period: The case of Neolithic variscite jewels from Western Europe (5th–3th millennium BC). The 13th International Conference on Particle Induced X-ray Emission (PIXE 2013) 318, Part A (jan): 149–56. https://doi.org/10.1016/j.nimb.2013.07.033.

Richerson, P. J., R. Boyd, and R. L. Bettinger. 2001. Was Agriculture Impossible during the Pleistocene but Mandatory during the Holocene? A Climate Change Hypothesis. *American Antiquity* 66 (3): 387–411.

Riede, F. 2011. Steps towards operationalizing an evolutionary archaeological definition of culture. In *Investigating Archaeological Cultures: Material Culture, Variability, and Transmission*, ed. by B. W. Roberts and M. V. Linden, pp. 245–70. New York: Spinger Verlag.

Rigaud, S. 2011. *La parure: Traceur de la géographie culturelle et des dynamiques de peuplement au passage Mésolithique-Néolithique en Europe*. Université Sciences et Technologies, Université de Bordeaux.

Rigaud, S. 2013. Les objets de Parure associés au dépôt funéraire mésolithique de Große Ofnet: Implications pour la compréhension de l'organisation sociale des dernières sociétés de chasseurs-cueilleurs du Jura Souabe. *Anthropozoologica* 48 (2): 207–230. http://www.bioone.org/doi/abs/10.5252/az2013n2a2.

Rigaud, S., S. Costamagno, J.-M. Pétillon, P. Chalard, V. Laroulandie, and M. Langlais. 2019. Settlement Dynamic and Beadwork: New Insights on Late Upper Paleolithic Craft Activities. *PaleoAnthropology* 2019: 137–55.

Rigaud, S., F. d'Errico, and M. Vanhaeren. 2015. Ornaments Reveal Resistance of North European Cultures to the Spread of Farming. *PLoS ONE* 10 (4): e0121166. https://doi.org/10.1371/journal.pone.0121166.

Rigaud, S., F. d'Errico, M. Vanhaeren, and X. Peñalber. 2014. A short-term, task-specific site: Epipalaeolithic settlement patterns inferred from marine shells found at Praileaitz I (Basque Country, Spain). *Journal of Archaeological Science* 41 (0): 666–78. https://doi.org/10.1016/j.jas.2013.10.009.

Rigaud, S., and I. Gutiérrez-Zugasti. 2016. Symbolism among the Last Hunter–Fisher–Gatherers in Northern Iberia: Personal Ornaments from El Mazo and El Toral III Mesolithic Shell Midden Sites. *Quaternary International* 407 (July): 131–44.

Rigaud, S., C. Manen, and I. García-Martínez de Lagrán. 2018. Symbols in motion: Flexible cultural boundaries and the fast spread of the Neolithic in the western Mediterranean. *PLOS ONE* 13 (5): e0196488. https://doi.org/10.1371/journal.pone.0196488.

Rigaud, S., M. Vanhaeren, A. Queffelec, G. Bourdon, and F. d'Errico. 2013. The Way We Wear Makes the Difference: Residue Analysis Applied to Mesolithic Personal Ornaments from Hohlenstein-Stadel (Germany). *Archaeological and Anthropological Sciences* 6 (2014): 133–144. https://doi.org/10.1007/s12520-013-0169-9.

Rindos, D. 1984. *The Origins of Agriculture: An Evolutionary Perspective*. Academic Press.

Rowley-Conwy, P., and R. Layton. 2011. Foraging and farming as niche construction: Stable and unstable adaptations. *Philosophical Transactions of the Royal Society B: Biological Sciences* 366 (1566): 849–62.

Sánchez-Quinto, F., H. Schroeder, O. Ramirez, M. C. Ávila-Arcos, M. Pybus, I. Olalde, A. M. V. Velazquez, et al. 2012. Genomic Affinities of Two 7,000-Year-Old Iberian Hunter-Gatherers. *Current Biology* 22 (16): 1494–99. https://doi.org/10.1016/j.cub.2012.06.005.

Sanders, J. M. 2002. Ethnic Boundaries and Identity in Plural Societies. *Annual Review of Sociology* 28: 327–57.

Shennan, S. J. 2002. *Genes, memes and human history. Darwinian archaeology and cultural evolution*. London: Thames and Hudson.

Shennan, S. 2011. Descent with modification and the archaeological record. *Philosophical Transactions of the Royal Society B: Biological Sciences* 366 (1567): 1070–79. https://doi.org/10.1098/rstb.2010.0380.

Shennan, S., S. S. Downey, A. Timpson, K. Edinborough, S. Colledge, T. Kerig, K. Manning, and M. G. Thomas. 2013. Regional population collapse followed initial agriculture booms in mid-Holocene Europe. *Nature Communications* 4 (1): 2486. https://doi.org/10.1038/ncomms3486.

Shennan, S. J., E. R. Crema, and T. Kerig. 2015. Isolation-by-distance, homophily, and "core" vs. "package" cultural evolution models in Neolithic Europe. *Evolution and Human Behavior* 2: 103–9. https://doi.org/10.1016/j.evolhumbehav.2014.09.006.

Skoglund, P., H. Malmström, A. Omrak, M. Raghavan, C. Valdiosera, T. Günther, P.Hall, et al. 2014. Genomic Diversity and Admixture Differs for Stone-Age Scandinavian Foragers and Farmers. *Science* 344 (6185): 747–50.

Soares, P., A. Achilli, O. Semino, W. Davies, V. Macaulay, H.-J. Bandelt, A. Torroni, and M. B. Richards. 2010. The Archaeogenetics of Europe. *Current Biology* 20 (4): R174–83. https://doi.org/10.1016/j.cub.2009.11.054.

Tartar, E., N. Reyssandier, F. Bon, and D. Liolios. 2006. Équipement de chasse, équipement domestique: une distinction efficace ? Réflexion sur la notion d'investissement technique dans les industries aurignaciennes. In *Normes techniques et pratiques sociales: de la simplicité des outillages pré- et protohistoriques*, ed. by L. Astruc, F. Bon, V. Léa, P.-Y. Milcent, S. Philibert, pp. 107–17. Éditions Antibes: APDCA.

Tehrani, J., and M. Collard. 2009. On the relationship between inter-individual cultural transmission and population-level cultural diversity: a case-study of weaving in Iranian tribal populations. *Evolution and Human Behavior* 30: 286–300.

Tresset, A., and J. D. Vigne. 2007. Substitution of species, techniques and symbols at the Mesolithic-Neolithic transition in Western Europe. *Proceedings of the British Academy* 144: 189–210.

Verswijver, G. 1986. Analyse comparative des parures Nahua: similitudes et différences. Musée d'Ethnographie de la Ville de Genève. *Bulletin Annuel* 29: 25–67.

Vigne, J.-D., I. Carrère, F. Briois, and J. Guilaine. 2011. The Early Process of Mammal Domestication in the Near East: New Evidence from the Pre-Neolithic and Pre-Pottery Neolithic in Cyprus. *Current Anthropology* 52 (S4): S255–71. https://doi.org/10.1086/659306.

Weninger, B., E. Alram-Stern, E. Bauer, L.Clare, U. Danzeglocke, O. Jöris, C. Kubatzki, G. Rollefson, H. Todorova, and T. van Andel. 2006. Climate forcing due to the 8200 cal yr BP event observed at Early Neolithic sites in the eastern Mediterranean. *Quaternary Research* 66: 401–20.

White, R. 2007. Systems of personal ornamentation in the early Upper Palaeolithic: methodological challenges and new observations. In *Rethinking the Human Revolution: New Behavioural & Biological Perspectives on the Origins and Dispersal of Modern Humans*, ed.by P. Mellars and C. Stringer, pp. 1–30. Cambridge: McDonald Institute for Archeological Research.

Wright, K. I. 2014. Domestication and inequality? Households, corporate groups and food processing tools at Neolithic Çatalhöyük. *Journal of Anthropological Archaeology* 33 (March): 1–33. https://doi.org/10.1016/j.jaa.2013.09.007.

Zvelebil, M. 2006. Mobility, contact, and exchange in the Baltic Sea basin. *Journal of Anthropological Archaeology* 25: 178–92.

Chapter 5

Genetic demography: What does it mean and how to interpret it, with a case study on the Neolithic transition

Michela Leonardi[1], Guido Barbujani[2], Andrea Manica[1]

Abstract

The present work describes the basic principles underlying demographic reconstructions from genetic data, and reviews the studies using such methods with respect to the Neolithic Demographic Transition. It is intended as a tool for scholars outside the field of population genetics (e.g., archaeologists, anthropologists, etc.) to better understand the significance and intrinsic limitations of genetic demography, and to help integrate its results within the broader context of the reconstruction of the human past.

THE PRINCIPLES OF GENETIC DEMOGRAPHY: NULL MODELS, GENETIC DRIFT AND EFFECTIVE POPULATION SIZE

Genetic demography is the branch of population genetics aiming at inferring changes in the size of one or more given populations from genetic data. It is widely used to reconstruct the demographic trajectories through time (e.g., Leonardi, Barbujani and Manica 2017; Miller, Manica and Amos 2018) or to test which demographic model within a set of explicit ones better fits the observed data (e.g., Leonardi et al. 2018; Vai et al. 2019).

It is important to highlight that, in population genetics, demography is always considered in terms of *effective population size* (N_e), which, as we will discuss, is not a proxy for census size.

---

[1] Department of Zoology, University of Cambridge, United Kingdom.
[2] Dipartimento di Scienze della Vita e Biotecnologie, Università di Ferrara, Italy.

© 2021, Kerns Verlag / https://doi.org/10.51315/9783935751377.005
Cite this article: Leonardi, M., G. Barbujani, and A. Manica. 2021. Genetic demography: What does it mean and how to interpret it, with a case study on the Neolithic transition. In *Ancient Connections in Eurasia*, ed. by H. Reyes-Centeno and K. Harvati, pp. 91-100. Tübingen: Kerns Verlag. ISBN: 978-3-935751-37-7.

This section will answer the question "what is effective population size?" by putting the concept within its historical context. This will help to highlight its importance in the field of population genetics and its limitations when borrowed by other fields.

Population genetics can be seen, in a broad way, as an attempt to create models as precise as possible of the way a population evolves, i.e. its genetic diversity is transmitted from a generation to the following one.

The basis for this process started at the beginning of the 20th century, when a (very unrealistic – but extremely useful) "null model" was created, defining how populations behave when no evolutionary pressure is acting (Hardy 1908; Weinberg 1908).

The so-called *Hardy-Weinberg equilibrium* shows how a population is in equilibrium, i.e., does not change its allele frequencies from one generation to the following one, when it meets the following assumptions:

- Sexual reproduction and diploidy (two copies of the genome in each cell);
- Non-overlapping generations;
- Random mating;
- (Infinitely) large population size;
- Negligible migration and mutation;
- Mortality and fertility independent from genotype.

This model shows that, for evolution to occur, at least one of the assumptions is not met. The next steps, therefore, were to find the way each of these assumptions influences the evolution of a population.

A few years later, Wright and Fisher expanded this model (developing the *Wright-Fisher model*) to estimate what happens to allele frequencies when only the assumption of infinitely large population size is not met (Fisher 1923; Wright 1931). In a finite population, at each generation the individuals may or may not leave descendants, who then in turn may or may not survive: this results in each generation being a random sample of the previous one. This effect of random sampling is called *genetic drift* and is stronger (i.e., causes broader changes between generations) in smaller population relative to larger ones.

Now it is possible to get back to our original question: effective population size is defined, somewhat tautologically, as "the size of a Wright-Fisher population experiencing the same genetic drift as the one under study" (Jobling et al. 2014) which, as we have seen, can also be translated into "the number of individuals contributing to the following generation" (Ayala 1982).

In a nutshell, it measures the effect of random sampling on the population analyzed: the larger the population, the smaller the expected changes between generations due to random sampling of gametes.

Effective population size can be estimated on the basis of many different types of genetic markers with different methods, reconstructing its variation through time. It must be highlighted, though, that such methods either tend to assume a single unstructured population evolving in isolation, or very simple effects of migration and other processes (e.g., just a few changes in population sizes through long lapses of time). It is important to consider that all the evidence discussed in this chapter is produced with methods of the first class, i.e., assuming negligible migration and geographic/cultural/social structure.

INTERPRETING EFFECTIVE POPULATION SIZE

It is very important to stress that estimates of effective population size hardly correspond to a something that, taken at face value, can be meaningful for other disciplines (Hawks 2008).

Effective population size can be regarded as a measure of the number of individuals actually contributing to the gene pool of the following generation. As a rule of thumb, in many cases it has been approximated as one-third of the census size, but there is no guarantee this assumption holds in every specific case. Therefore, census size is only one of the determinants of effective population size, together with a host of other factors such as gene flow (i.e., immigration/emigration), geographic structuring, age structure, mating/marriage patterns, sex ratio, breeding practices (for domesticates), social structure (for humans), etc. In the absence of genetic data from other populations, as well as additional information from other lines of evidence such as archaeology and historical records, it is impossible to disentangle the effect of these factors.

As an example, a population that has some form of geographic, social or cultural structuring (i.e., where individuals are more likely to have offspring within a specific subset of the population), will tend to have a larger effective size than a population with the same census size but where mating is completely random (Waples 2010). The reason is that, in such a scenario, each sub-population is relatively isolated from the others, and thus affected by drift in an independent manner; this effect is compounded by the fact that each sub-population is also smaller in size than the whole metapopulation, leading to stronger drift. The result is that sub-populations will end up differing from each other in their allele frequencies, potentially even developing private variants. In a random mating scenario, this level of genetic variability would require a much larger population size (Charlesworth, Charlesworth, and Barton 2003).

Another important point is that from a genetic point of view it is not possible to define the area inhabited by a population and its density, which add another level of complexity when trying to link population size with census size. There have been attempts at solving this problem by comparing effective population size with the potential range of species reconstructed based on climatic data. While there seems to be some level of correlation between the two in some cases (Lorenzen et al. 2011), this is not always true (Miller et al. 2021).

For all the reasons mentioned above, when we observe an increase in effective population size through time in our data it may be the result of many different scenarios. In the following list, as an example, we will detail some such possibilities both as general concepts and in a few situations that may occur in a human population (in italic).

- Increased population density when occupying the same geographic area: *following a climatic amelioration, the environmental productivity increases and the occupied region can sustain more individuals; or, the development of new technologies allows better use of the available resources.*
- An increase in fertility: *the transition from a nomadic to a sedentary way of life changes the group behavior, allowing an increase in the number of children that a family can sustain at a given time.*
- An increase in the geographic area inhabitable: *climatic changes give access to a previously unoccupied region.*
- Beginning or increase in immigration from one or more populations, genetically different from the original one: *a military conquest; or, the development of new trading routes; or, immigration of a new population in the area.*
- A change in marriage rules: *after a military conquest foreign men tend to marry local women; or, because of new commercial connections there is an increase in marriages between the two populations involved, to strengthen such economical relationships.*
- Any other situation leading to an increase of the genetic diversity within the population, including different combinations of the proposed scenarios.

Similarly, below we present some examples of processes that lead to a decrease in effective population size:

- Strong reduction in census size (in genetic terms defined as a *bottleneck*): *an epidemic; or, a war; or, a decrease in food availabilty.*
- A small group from the parental population moving away from the latter: *the occupation of a new area by a subset of individuals* (in genetic terms defined *founder event*);
- Reduction or interruption of genetic exchange with other populations (i.e., *isolation*): *degradation and finally unavailability of a previously developed route or trade network.*
- Population fragmentation (e.g., because of the emergence of geographical, cultural or social barriers within it): *the development of a caste-like social system that excludes marriage between different social classes.*
- Increase in mortality: *increase of child mortality due to famine.*
- Decrease in fertility: *the needs of more frequent or longer migrations reduces the number of young children that the group can*

carry at the same time; or, migrating in a harsher climate increases the time to reach sexual maturation in women.
- Reduction or geographical shift of the area environmentally habitable by the population (leading to the survival of only of a part of it): *a climatic change reducing the productivity of the region inhabited (e.g., Sahara)*.
- Increase of marriage between relatives: *to keep power or wealth*.
- Reduction in the number of individuals of a specific sex (there is a stronger effect if this affects females): *a war*.
- For domesticates: *starting or change in breeding practices*.
- Any other situation leading to a decrease of the genetic diversity within the population including combinations of the scenarios above.

CASE STUDY: GENETIC DEMOGRAPHY AND THE NEOLITHIC DEMOGRAPHIC TRANSITION

The advent of the Neolithic, i.e. the transition from foraging to farming, led to an increase in population density in many different regions of the world (Armelagos, Goodman, and Jacobs 1991; Bellwood et al. 2007; Kılınç et al. 2016; J.-P. Bocquet-Appel 2011; J. P. Bocquet-Appel and Bar-Yosef 2008).

From a genetic point of view, many studies have approached the subject by reconstructing the genetic demography in the present and through time. It is well established that the effective population size significantly differs between modern-day food producers on the one hand and foragers on the other, with the latter showing much smaller values (Excoffier and Schneider 1999; Destro-Bisol et al. 2004; Pilkington et al. 2008; Aimé et al. 2014; Patin et al. 2014; Leonardi, Barbujani, and Manica 2017; Gopalan et al. 2019).

Several studies used different portions of the genome to reconstruct the demographic profile in populations with contrasting subsistence strategies to date the onset of these differences. The expectation was that food producers and foragers shared similar trajectories until more or less the Neolithic, when the former started increasing in numbers while the latter did not (Menozzi, Piazza, and Cavalli-Sforza 1978; Sokal, Oden, and Wilson 1991). However, contrary to this hypothesis, a large number of studies suggest a significant difference between the two groups and an increase in effective population size in farmers starting before the Neolithic (Leonardi, Barbujani, and Manica 2017; Atkinson, Gray, and Drummond 2008; Batini et al. 2015, 2011; Maisano Delser et al. 2017; Chaix et al. 2008; Zheng et al. 2011, 2012; Aimé et al. 2013; Patin et al. 2014; Miller, Manica, and Amos 2018). Since all populations relied on a hunting-gathering lifestyle before the Neolithic transition, this finding calls for an explanation.

A first possibility is that the observed evidence could be due to climate: an uneven distribution of the natural resources and/or the climatic amelioration following the Last Glacial Maximum may have allowed some environments to sustain larger groups (Bar-Yosef 1998; Berger and Guilaine 2009). For this reason, both the differences between lifestyles and the early increase in the effective population size in farmers could be the result of the ancestors of modern-day food producers living in more productive environments than the ancestors of modern foragers. Such a more favorable climate could also have facilitated the development of food production, increasing the environmental productivity for foragers and allowing them to sustain even larger groups.

Was that the case? To answer this question, our recent paper (Leonardi, Barbujani, and Manica 2017) analyzed the trajectories of effective population sizes of present-day foragers and food producers from Sub-Saharan Africa, South-Eastern Asia and Siberia over the last twenty thousand years. Changes in effective populations sizes were then compared with estimates of Net Primary Productivity through time. In all three regions, food producers systematically show higher numbers of effective individuals than foragers, even after correcting for environmental productivity. Furthermore, the trajectories also indicate higher effective growth rates in the farmers, and this difference can only be attributed to farmers living in climatically more favourable regions for Siberian populations (but not for Sub-Saharan Africa and South-East Asia).

Another possibility is that we are observing the result of behavioral and social differences among Paleolithic hunter-gatherers. Present-day foragers exhibit large variability in term of complex behaviors that may influence their genetic diversity (and hence the estimation of effective population size), e.g., sedentism, storage activity and social stratification (Rowley-Conwy 2001). Those three aspects are all typical of food producers, and appear to be all linked to stantiality. Indeed, it has been suggested that the transition to farming happened in foraging populations that were already sedentary, because they would have had the resources to sustain themselves while experimenting with agriculture and the continuity to check the cultivations as they developed (Sauer 1952). It has also been shown that, in modern-day hunter-gatherers, a larger ratio of population size over ecological productivity is positively correlated to a sedentary lifestyle, storage of goods and hierarchy, which means that populations that exhibit such behaviors are likely to have larger effective population size than other groups (Rowley-Conwy 2001).

The observed differences in effective population size are then coherent with a scenario where the Paleolithic ancestors of modern food producers were more sedentary/socially stratified/interconnected/etc. than the contemporary ancestors of modern-day foragers.

It is also possible that food producers originated from populations with larger effective size as both the main factors influencing it (higher population density or higher rates of gene flow and so cultural connectivity with neighboring groups) could facilitate technical innovation, possibly leading to an improvement in subsistence technologies. Indeed, a link between higher effective size and increase of cultural complexity has been observed in different contexts (Powell, Shennan, and Thomas 2009).

It is important to stress that, as discussed above, the methods used to obtain the evidence presented rely on the assumption of a single population with negligible migration and population structure (e.g., geographic, social, cultural). In contrast, Neolithization involved the spread of human groups from a different region and subsequent mixing with local foragers was more than likely (e.g., Lazaridis et al. 2014; Prendergast et al. 2019). Whether, and to what extent, admixture between expanding early farmers and local hunter-gatherers affected the results of retrospective studies is difficult to tell. However, the observed gradual changes in effective population size and the consistent pattern all over the world do not suggest that admixture, which likely took place at different rates in different areas, has been the main cause of this phenomenon.

CONCLUSIONS

In short, changes in effective population sizes inform us that some process has been affecting the genetic diversity of the population analyzed, and identifying such process(es) should be done carefully and in the light of the archaeological/historical context.

The terms *demography* and *effective population size* suggest a variation in census size as the main reason to explain such evidence. However, and to the contrary, the results of genetic demographic analyses may be linked to a variety of processes. Luckily in many cases, archaeology, anthropology and other sister disciplines have the data and methods to test different possibilities and find the correct one. Modern Bayesian methods of demographic inference associated with simulation studies (Beaumont, Zhang, and Balding 2002) allow one to model in greater detail demographic changes, selecting the model best fitting the data among various options.

For example, immigration and introgression from a different population may be tested through the analysis of anthropological, isotopic and cultural data; the development (or collapse) of a trade network can be shown by studying the distribution through time of the sources of raw materials; social stratification can be suggested by looking at mortuary practices; the distribution of archaeological sites through space and time could be informative about population density, etc.

The reconstructions of genetic demography should therefore not be considered as providing definitive answers, but rather as a starting point for debate. A debate that can only be resolved through interdisciplinarity.

ACKNOWLEDGMENTS

ML and AM have been funded by the ERC Consolidator Grant 647787 "LocalAdaptation". All Authors thank the two anonymous reviewers whose comments helped improving the quality of this manuscript.

REFERENCES

Aimé, C., G. Laval, E. Patin, P. Verdu, L. Segurel, R. Chaix, T. Hegay, L. Quintana-Murci, E. Heyer, and F. Austerlitz. 2013. Human Genetic Data Reveal Contrasting Demographic Patterns between Sedentary and Nomadic Populations That Predate the Emergence of Farming. *Molecular Biology and Evolution* 30 (12): 2629–44.

Aimé, C., P. Verdu, L. Ségurel, B. Martinez-Cruz, T. Hegay, E. Heyer, and F. Austerlitz. 2014. Microsatellite Data Show Recent Demographic Expansions in Sedentary but Not in Nomadic Human Populations in Africa and Eurasia. *European Journal of Human Genetics* 22 (10): 1201–7.

Armelagos, G. J., A. H. Goodman, and K. H. Jacobs. 1991. The Origins of Agriculture: Population Growth during a Period of Declining Health. *Population and Environment* 13 (1): 9–22.

Atkinson, Q. D., R. D. Gray, and A. J. Drummond. 2008. MtDNA Variation Predicts Population Size in Humans and Reveals a Major Southern Asian Chapter in Human Prehistory. *Molecular Biology and Evolution* 25 (2): 468–74.

Ayala, F. J. 1982. *Population and Evolutionary Genetics: A Primer*. The Benjamin/Cummings Publishing Company.

Bar-Yosef, O. 1998. The Natufian Culture in the Levant, Threshold to the Origins of Agriculture. *Evolutionary Anthropology: Issues, News, and Reviews* 6 (5): 159–77. https://doi.org/10.1002/(SICI)1520-6505(1998)6:5<159::AID-EVAN4>3.0.CO;2-7.

Batini, C., P. Hallast, D. Zadik, P. M. Delser, A. Benazzo, S. Ghirotto, E. Arroyo-Pardo, et al. 2015. Large-Scale Recent Expansion of European Patrilineages Shown by Population Resequencing. Nature Communications 6: 7152. https://doi.org/10.1038/ncomms8152.

Batini, C., J. Lopes, D. M. Behar, F. Calafell, L. B. Jorde, L. van der Veen, L. Lluis Quintana-Murci, G. Spedini, G. Destro-Bisol, and D. Comas. 2011. Insights into the Demographic History of African Pygmies from Complete Mitochondrial Genomes. *Molecular Biology and Evolution* 28 (2): 1099–1110. https://doi.org/10.1093/molbev/msq294.

Beaumont, M. A., W. Zhang, and D. J. Balding. 2002. Approximate Bayesian Computation in Population Genetics. *Genetics* 162 (4): 2025–2035.

Bellwood, P., C. Gamble, S. a. Le Blanc, M. Pluciennik, M. Richards, and J. E. Terrell. 2007. First Farmers: The Origins of Agricultural Societies. *Cambridge Archaeological Journal* 17 (01): 87. https://doi.org/10.1017/S0959774307000078.

Berger, J. F., and J. Guilaine. 2009. The 8200 Cal BP Abrupt Environmental Change and the Neolithic Transition: A Mediterranean Perspective. *Quaternary International* 200 (1–2): 33–49.

Bocquet-Appel, J.-P. 2011. When the World's Population Took off: The Springboard of the Neolithic Demographic Transition. *Science* 333 (6042): 560–61. https://doi.org/10.1126/science.1208880.

Bocquet-Appel, J.- P., and O. Bar-Yosef, eds. 2008. *The Neolithic Demographic Transition and Its Consequences*. Springer. https://doi.org/10.1007/978-1-4020-8539-0.

Chaix, R., F. Austerlitz, T. Hegay, L. Quintana-Murci, and E. Heyer. 2008. Genetic Traces of East-to-West Human Expansion Waves in Eurasia. *American Journal of Physical Anthropology* 136 (3): 309–17. https://doi.org/10.1002/ajpa.20813.

Charlesworth, B., D. Charlesworth, and N. H. Barton. 2003. The Effects of Genetic and Geographic Structure on Neutral Variation. *Annual Review of Ecology, Evolution, and Systematics* 34: 99–125. https://doi.org/10.1146/annurev.ecolsys.34.011802.132359.

Destro-Bisol, G., F. Donati, V. Coia, I. Boschi, F. Verginelli, A. Caglià, S. Tofanelli, G. Spedini, and C. Capelli. 2004. Variation of Female and Male Lineages in Sub-Saharan Populations: The Importance of Sociocultural Factors. Molecular Biology and Evolution 21 (9): 1673–82. https://doi.org/10.1093/molbev/msh186.

Excoffier, L., and S Schneider. 1999. Why Hunter-Gatherer Populations Do Not Show Signs of Pleistocene Demographic Expansions. Proceedings of the National Academy of Sciences of the United States of America 96 (19): 10597–602.

Fisher, R. A. 1923. On the Dominance Ratio. Proceedings of the Royal Society of Edinburgh 42: 321–41.

Gopalan, S., R. E. W. Berl, G. Belbin, C. Gignoux, M. W. Feldman, B. S. Hewlett, and B. M. Henn. 2019. Hunter-Gatherer Genomes Reveal Diverse Demographic Trajectories Following the Rise of Farming in East Africa. BioRxiv. https://doi.org/10.1101/517730.

Hardy, G. H. 1908. Mendelian Proportions in a Mixed Population. Science. https://doi.org/10.1126/science.28.706.49.

Hawks, J. 2008. From Genes to Numbers: Effective Population Sizes in Human Evolution. In Recent Advances in Palaeodemography: Data, Techniques, Patterns, ed. by J.-P- Bocquet-Appel, pp. 9–30. Springer Science+Business Media. https://doi.org/10.1007/978-1-4020-6424-1_1.

Jobling, M., E. J. Hollox, M. E. Hurles, T. Kivisild, and C. Tyler-Smith. 2014. *Human Evolutionary Genetics*. 2nd edition. Garland Science. https://www.routledge.com/Human-Evolutionary-Genetics/Jobling-Hollox-Kivisild-Tyler-Smith/p/book/9780815341482.

Kılınç, G. M., A. Omrak, F. Özer, T. Günther, A. M. Büyükkarakaya, E. Bıçakçı, D. Baird, et al. 2016. The Demographic Development of the First Farmers in Anatolia. Current Biology 0 (0): 137–40. https://doi.org/10.1016/j.cub.2016.07.057.

Lazaridis, I., N. Patterson, A. Mittnik, G. Renaud, S. Mallick, K. Kirsanow, P. H. Sudmant, et al. 2014. Ancient Human Genomes Suggest Three Ancestral Populations for Present-Day Europeans. Nature 513 (7518): 409–13. https://doi.org/10.1038/nature13673.

Leonardi, M., G. Barbujani, and A. Manica. 2017. An Earlier Revolution: Genetic and Genomic Analyses Reveal Pre-Existing Cultural Differences Leading to Neolithization. Scientific Reports 7 (1): 3525. https://doi.org/10.1038/s41598-017-03717-6.

Leonardi, M., A. Sandionigi, A. Conzato, S. Vai, M. Lari, F. Tassi, S. Ghirotto, D. Caramelli, and G. Barbujani. 2018. The Female Ancestor's Tale: Long-Term Matrilineal Continuity in a Nonisolated Region of Tuscany. *American Journal of Physical Anthropology* 167 (3): 497–506. https://doi.org/10.1002/ajpa.23679.

Lorenzen, E. D., D. Nogués-Bravo, L. Orlando, J. Weinstock, J. Binladen, K. A. Marske, A. Ugan, et al. 2011. Species-Specific Responses of Late Quaternary Megafauna to Climate and Humans. *Nature* 479 (7373): 359–64. https://doi.org/10.1038/nature10574.

Maisano Delser, P., R. Neumann, S. Ballereau, P. Hallast, C. Batini, D. Zadik, and M. A. Jobling. 2017. Signatures of Human European Palaeolithic Expansion Shown by Resequencing of Non-Recombining X-Chromosome Segments. *European Journal of Human Genetics* 25 (4): 485–492. https://doi.org/10.1038/ejhg.2016.207.

Menozzi, P., A. Piazza, and L. L. Cavalli-Sforza. 1978. Synthetic Maps of Human Gene Frequencies in Europeans. *Science* 201 (4358): 786–92.

Miller, E. F., A. Manica, and W. Amos. 2018. Global Demographic History of Human Populations Inferred from Whole Mitochondrial Genomes. *Royal Society Open Science* 5: 180543. https://doi.org/10.1098/rsos.180543.

Miller, E. F., R. E. Green, A. Balmford, P. Maisano Delser, R. Beyer, M. Somveille, M. Leonardi, W. Amos, and A. Manica. 2021. Bayesian Skyline Plots disagree with range size changes based on Species Distribution Models for Holarctic birds. *Molecular Ecology*. Accepted Author Manuscript. https://doi.org/10.1111/mec.16032.

Patin, E., K. J. Siddle, G. Laval, H. Quach, C. Harmant, N. Becker, A. Froment, et al. 2014. The Impact of Agricultural Emergence on the Genetic History of African Rainforest Hunter-Gatherers and Agriculturalists. *Nature Communications* 5: 3163. https://doi.org/10.1038/ncomms4163.

Pilkington, M. M., J. A. Wilder, F. L. Mendez, M. P. Cox, A. Woerner, T. Angui, S. Kingan, et al. 2008. Contrasting Signatures of Population Growth for Mitochondrial DNA and Y Chromosomes among Human Populations in Africa. *Molecular Biology and Evolution* 25 (3): 517–25. https://doi.org/10.1093/molbev/msm279.

Powell, A., S. Shennan, and M. G. Thomas. 2009. Late Pleistocene Demography and the Appearance of Modern Human Behavior. *Science* 324 (5932): 1298–1301. https://doi.org/10.1126/science.1170165.

Prendergast, M. E., M. Lipson, E. A. Sawchuk, I. Olalde, C. A. Ogola, N. Rohland, K. A. Sirak, et al. 2019. Ancient DNA Reveals a Multistep Spread of the First Herders into Sub-Saharan Africa. *Science* 365 (6448): eaaw6275. https://doi.org/10.1126/science.aaw6275.

Rowley-Conwy, P. 2001. Time, Change and the Archaeology of Hunter-Gatherers : How Original Is the 'Original Affluent Society'? In *Hunter-Gatherers: An Interdisciplinary Perspective*, ed. by C. Panter-Brick, R. H. Layton, and P. Rowley-Conwy, pp. 39–72. Cambridge: Cambridge University Press.

Sauer, C. O. 1952. *Agricultural Origins and Dispersals*. Cambridge (MA): American Geographical Society.

Sokal, R. R., N. L. Oden, and C. Wilson. 1991. Genetic Evidence for the Spread of Agriculture in Europe by Demic Diffusion. *Nature* 351: 143–145. https://doi.org/10.1038/351143a0.

Vai, S., A. Brunelli, A. Modi, F. Tassi, C. Vergata, E. Pilli, M. Lari, et al. 2019. A Genetic Perspective on Longobard-Era Migrations. *European Journal of Human Genetics* 27 (4): 647–56. https://doi.org/10.1038/s41431-018-0319-8.

Waples, R. S. 2010. Spatial-Temporal Stratifications in Natural Populations and How They Affect Understanding and Estimation of Effective Population Size. *Molecular Ecology Resources* 10 (5): 785–96. https://doi.org/10.1111/j.1755-0998.2010.02876.x.

Weinberg, W. 1908. Über Den Nachweis Der Vererbung Beim Menschen. *Jahreshefte des Vereins für Vaterländische Naturkunde in Württemberg*: 64: 368–82.

Wright, S. 1931. Evolution in Mendelian Populations. *Genetics* 16: 97–159.

Zheng, H.-X., S. Yan, Z.-D. Qin, and L. Jin. 2012. MtDNA Analysis of Global Populations Support That Major Population Expansions Began before Neolithic Time. *Scientific Reports* 2: 745. https://doi.org/10.1038/srep00745.

Zheng, H.-X., S. Yan, Z.-D. Qin, Y. Wang, J.-Z. Tan, H. Li, and L. Jin. 2011. Major Population Expansion of East Asians Began before Neolithic Time: Evidence of MtDNA Genomes. *PloS One* 6 (10): e25835. https://doi.org/10.1371/journal.pone.0025835.

Chapter 6

Statistical methods for kinship inference amongst ancient individuals

Andaine Seguin-Orlando[1]

Abstract

The identification of close relatives is central to forensic sciences and to genetic association studies, in which spurious signals can be obtained if genetic structure is not taken into account. Identifying related individuals is also essential in archaeological studies to elucidate funerary practices, as well as to obtain a deeper understanding of past family structures and social behaviors in the absence of written records. In the past decade, following the advent of high-throughput DNA sequencing, many statistical methods have been developed to calculate kinship coefficients from genome-wide data. However, these methods are inappropriate when DNA is sequenced at insufficient depth-of-coverage, presenting high levels of post-mortem damage, as is commonly observed with ancient molecules. These methods also generally require the presence of a reference panel, which cannot be accessed in the vast majority of paleogenomic studies. Here, I review the different approaches available for inferring relatedness, focusing on those compatible with the idiosyncrasies of ancient genomic data. I then present some of the key studies taking advantage of these analytical tools, ranging from simple sample curation to addressing long-standing archaeological debates on the emergence of the nuclear family and on the role of biological kinship in past societies.

INTRODUCTION: WHY IS RELATEDNESS INFERENCE IMPORTANT?

Many genetic tests have become commercially available in the last decade, especially for those interested in their genealogical history and/or deeper ancestry. For less than a hundred dollars and a saliva sample, ancestry genetic testing companies provide their customers with relatedness estimates, compared against the data of whomever agreed to share

---

[1] Centre for Anthropobiology and Genomics of Toulouse, UMR 5288, CNRS - Université Paul Sabatier Toulouse III, Toulouse, France.

© 2021, Kerns Verlag / https://doi.org/10.51315/9783935751377.006
Cite this article: Seguin-Orlando, A. 2021. Statistical methods for kinship inference amongst ancient individuals. In *Ancient Connections in Eurasia*, ed. by H. Reyes-Centeno and K. Harvati, pp. 101-127. Tübingen: Kerns Verlag. ISBN: 978-3-935751-37-7.

genotyping information. Despite their success on a global scale, genetic ancestry kits have also raised legitimate concerns on ethics and data protection (Kennett et al. 2018).

Reliable methods for relatedness inference were developed for applications in a number of scientific fields. For example, in forensics, they are part of routine paternity testing, as well as for victim identification in cases where confident attribution of fragmentary remains to a single individual is not possible (Brenner and Weir 2003; Fernandes et al. 2017). In population genetics and genome-wide association studies, related individuals are identified in populations and cohorts and excluded from analyses since they do not represent statistically independent samples. In medical genetics, the presence of closely related individuals in association studies inflates rates of false positives; therefore, relatedness among individuals must always be tested beforehand. In practice, screening human panels for cryptic relatedness has become a routine quality control, as it is not rare that datasets contain unreported familial relationships as close as first degree (e.g., siblings), or even data from identical individuals (Stevens et al. 2011). Beyond medical genetics, inferring relatedness is crucial to the conservation of endangered species, as mating between closely related individuals leads to inbreeding. Finally, deciphering the genetic relatedness between individuals is key for understanding past human societies, as this information could provide major insights into social structures, traditions, cultures and behavior. For example, it has often been suggested that kindred played an important role in structuring past societies, but only the combination of archaeological evidence and genetic proof of biological relatedness can confirm this hypothesis.

In this chapter, I will summarize the different methods that are available today and make use of genome-wide sequence data to infer relatedness, with a focus on those tailored to ancient DNA data (Part I). I will then illustrate the wide range of applications that these methods have in archaeological science (Part II). The two parts may be read independently, depending on the readers' own interests.

PART I: DIFFERENT METHODS FOR DIFFERENT DATA

In the following, I survey the different methods available to infer kinship between past individuals, starting from the simple amplification by Polymerase Chain Reaction (PCR) of uniparental markers, to the most elaborate statistical processing of whole genome data. This part of the chapter will review the basics of pedigrees and kinship coefficients. It will also dive a bit deeper into the caveats of ancient DNA data, such as the low amount of available DNA molecules and the lack of contemporaneous reference panels, as well as the potential presence of population structure or the possibly high levels of inbreeding, which make it necessary to choose between alternative statistical methods that can account for these specificities.

Relatedness inference using genetic markers: Mitochondrial and Y-haplogroups, autosomal STR

Until recently, past potential kinship was mainly assessed through two types of uniparental markers: mitochondrial DNA and the non-recombining region of the Y chromosome. Despite its major impact in first-generation ancient DNA studies (Gerstenberger et al. 1999; Rogaev et al. 2009), as well as in forensic science with the example of paternity testing, this approach faces crucial pitfalls. Mitochondrial DNA is present in numerous copies per cell (which facilitates lab work), is maternally inherited and helps retrace maternal lineages. The Y-chromosome is paternally inherited, thus providing evolutionary and genealogical information on paternal lineages. Restricting the analyses to unilineal markers has two major consequences. First, as most of the mitochondrial and Y-chromosome haplotypes are widely distributed in populations, they can only be used reliably to exclude close relatedness, such as father-son and mother-offspring relationships from Y-chromosome and mitochondrial DNA data, respectively. Second, the Y chromosome only provides information on the unique lineage composed exclusively of the father at each generation and cannot elucidate father-daughter relationships. Reciprocally, mitochondrial DNA is only informative about maternal lineages. Looking back 10 generations ago in the genealogy of a given individual, those uniparental markers would give access to the genetic background of only two ancestors, and not the more than one thousand that in fact existed (in theory $2^{10}=1,024$, assuming unrelated ancestors). This contrasts to autosomal data, which can provide information transmitted by a larger number of ancestors (although not all).

As the quantity and quality of DNA samples available to forensic and ancient DNA laboratories may be very similar, any methodological development implemented in one field can benefit the other. For example, the gold standard for paternity testing in criminal or mass murder victim identification has been, from the 1990s onwards, the use of Short Tandem Repeats (STR), also known as microsatellites. In this approach, selected loci spread across the Y chromosome (or the autosomes), are amplified through PCR, and genotyped for the number of short motif repeats they carry. As these STR loci are multiallelic, characterizing only a few of them can help draw reliable conclusions on close familial relationships or consanguinity. For instance, the tests approved for paternity testing or offender identification are based on the amplification of 20 (US Combined DNA Index System) or 15 (Extended European Standard System Set) STR loci only. They show sufficient statistical power for elucidating parent-offspring relationships and the majority of sibling pairs, as well as for identifying historical characters. Famous examples are provided by the eight members of the House of Konigsfeld, Germany, spanning seven generations (1546-1749 AD, Gerstenberger 1999), or the last Russian Emperor Nicholas II, whose children have been identified and confirmed dead 91 years after they were murdered during the Russian Civil War (Rogaev et al. 2009).

Wider kinship, such as first cousins, cannot be reliably assessed on the basis of STR test profiles alone. In those cases, it becomes necessary to genotype a third individual who could be a closer relative to both individuals under investigation (Nothnagel et al. 2010). In addition, STRs have proven hard to genotype from degraded DNA extracted from ancient remains. Some of the targeted loci indeed span over more than 250 base pairs (bp, Gerstenberger et al. 1999), whereas the average fragment length for ancient DNA lies around 50 bp. Therefore, the success of STR amplification is generally very limited (Deguilloux et al. 2013) and, at best, only partial profiles can be obtained (Russo et al. 2016; Haak et al. 2008; Cui et al. 2015), which rules out this approach for most archaeological assemblages.

The development of molecular and bioinformatic tools tailored to ancient remains, as well as the rapidly growing High-Throughput Sequencing (HTS) technologies, made it possible to publish the first complete genome of a past human individual (Rasmussen et al. 2010). This major achievement opened the door to elucidating kinship between past humans not solely based on the isolated markers just mentioned, but on genome-scale data, even though genealogy and genetic relatedness are not synonymous, as described below.

From pedigree to relatedness inference, and back

The interest of the general public for commercial DNA testing (e.g., through the company *23andMe®*, etc.) has enhanced the confusion between genealogy and genetic relatedness, and between genetic data and culture. We should emphasize that genealogy and genetic relatedness cannot be superimposed (Reich 2018). The genealogy of one individual, as can be redrawn from familial and historical records, groups together all their known ascendants. In contrast, as a result of the random nature of DNA recombination, the genome of some of our numerous genealogical ascendants certainly did not contribute to any of our DNA. As a consequence, even the most extensive and accurate whole-genome analysis would not be able to redraw some of the family links, simply because the descendant did not inherit any fraction of the genome of some of his/her distant ancestors.

If two individuals are genetically related, they will not only share a recent common ancestor, as genealogic relatives do, but they would also have co-inherited a portion from the genome of that ancestor. If we consider two individuals that are the Nth generation descendants of one given couple, they will share an allele identical-by-descent (IBD) if this allele has been copied and transmitted to the two descendants over 2^N consecutive meiosis events. The probability of a copy to be transmitted over each generation equals 0.5 (1 chance over 2 chromosomes). Therefore the probability that the allele is inherited by each of the two N^{th} generation descendants under study is 2^{1-2N}. Following this, after only N=4 generations, the probability of the copy being inherited through both lineages appears less than 1% (~0.78%). When family trees (or pedigrees) are known, it is relatively straightforward to deduce the probability of any of the genotypes present in any pair of individuals. But the reciprocal is not true: finding the probability of an inferred relatedness knowing the genotypes, or sometimes even partial genotyping data or likelihood of genotypes, can be challenging. While it is not possible to track IBD alleles directly, one can instead leverage the observed genotypes to identify alleles identical-by-state (IBS). Those are used to obtain probabilities of the IBD status of the alleles, and then to make statistical inferences about the degree of relatedness. In the following chapter section, we will review the different coefficients expressing these degrees of relatedness, how they are calculated, and which conclusions can be drawn from their values.

How relatedness degrees and coefficients should be understood

Table 1 provides a summary of different familial relationships and their corresponding relatedness. The relatedness between two individuals can be represented by the shortest lineage path, or the number of parent-offspring steps, from one individual to the other. In the pedigree presented as an example in Figure 1A, **L** is the child of **F**, therefore the lineage path between them is one, over one single line of descent. In contrast, **I** and **L** are first cousins and can be linked by two lineage paths of length four, over two lines of descent (**I-D-A-F-L** and **I-D-B-F-L**).

The kinship coefficient (θ) quantifies the number of generation steps that separate the individuals of the pair and formalizes the relatedness between two individuals. In the pedigree presented in Figure 1A, **I** and **L** have both **A** and **B** as most recent common ancestors. Their kinship coefficient can be expressed as:

$$\theta_{(I,L)} = (1+f\mathbf{A})/2^{g\mathbf{A}+1} + (1+f\mathbf{B})/2^{g\mathbf{B}+1}$$

where g**X** represents the number of parent-offspring steps between **I** and **L** via **X** (**X** standing for either **A** or **B** in our case). f**X** is the inbreeding coefficient of **X**, corresponding in fact to the kinship coefficient of his/her parents. In our case, g**A**=g**B**=4. Assuming that **A** and **B** are unrelated (f**A**=f**B**=0), we obtain the following:

$$\theta_{(I,L)} = 2 \times (1/2^5) = 1/16 \text{ (i.e., 6.25\%).}$$

An allele from individual **I** has one in 16 chances of being descended from the same parental allele as an allele from **L**. In the eventuality of **I** and **L** being inbred, for instance if **A** and **B** were first cousins ($\theta_{(A,B)}=1/16$), we would have obtained a higher value of θ:

$$\theta_{(I,L)} = 2 \times ((1+0{,}0625)/2^5) = 17/256 \text{ (i.e., 6.64\%).}$$

In the context of ancient individuals, the pedigree is most likely to be unknown, and the calculations above can not be applied. A kinship coefficient is first calculated based on genomic data, and the most probable degree of relatedness between individuals is then deduced from this coefficient. The kinship coefficient θ can be defined as the probability that a randomly chosen allele in an individual is IBD to a randomly selected homologous allele in the second individual. For a pair of diploid individuals, there are 15 possible IBD states (Fig. 1B). It is generally not possible (but also not necessary in many cases) to distinguish between maternal or paternal alleles. Therefore, the possible different IBD sharing patterns for any given locus are condensed into nine and their probabilities are called Jacquard coefficients (Jacquard 1974). The frequencies of each of these nine distinct patterns estimate the common ancestry between the two individuals and their inbreeding level. When neither individual is expected to show significant level of inbreeding, IBD states can be further reduced to three (IBD=0, IBD=1 or IBD=2), with probabilities of IBD=0, IBD=1 and IBD=2 referred to as k_0, k_1 and k_2, respectively. In the case of no inbreeding (which is assumed in the majority of studies), θ can be expressed as follows:

$$\theta = k_1/4 + k_2/2.$$

In practice, IBD-segment based methods calculate k_1 and k_2 as the cumulative genetic length of the IBD=1 and IBD=2 segments, respectively, divided by the total genetic length of the genome. Relying solely

Fig. 1.
A: Example of a pedigree.
B: The identity states for individuals 1 and 2. They are lined up according to their nine condensed states and grouped according their three possible IBD sharing fraction. Points represent alleles and lines represent alleles that are identical.
C: Configurations expected for siblings. Top: On average for a sibling pair, 50% of the chromosome segments are expected to be Identical By Descent (IBD). One of the IBD segments is highlighted in grey. Bottom: When only low-coverage data are available, analyses benefit from incorporating genotypes likelihoods (GL). The 10 possible genotypes and their corresponding likelihoods GL0 to GL9 are presented in the table.

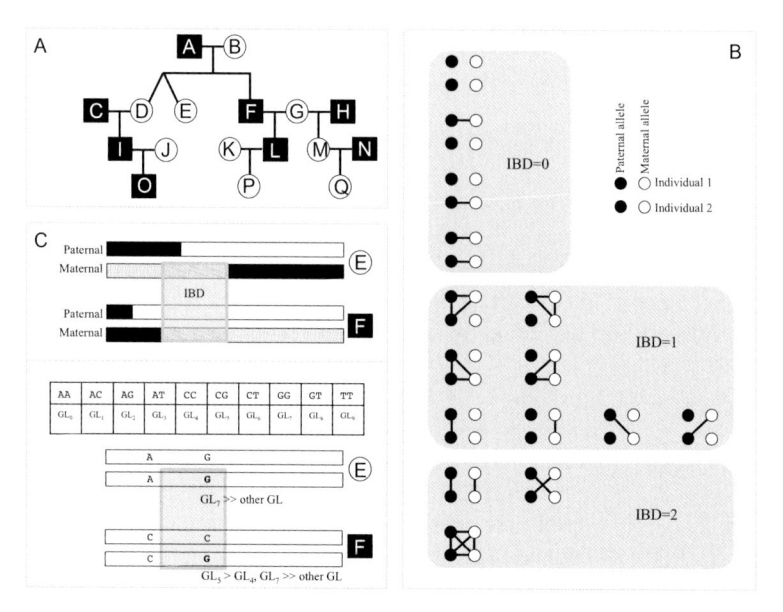

Table 1.
Different relatedness estimators.
Not all possible configurations for each degree of relatedness are mentioned.
Ex: example of a pair from Figure 1 with this degree of relatedness.
G: Number of generations or parent-offspring steps that separate the individuals of the pair.
L: Number of lines of descent the individuals of the pair belong to.
k_i: probability of IBD=i.
θ: kinship coefficient.
r: coefficient of relatedness.

Configuration	Ex	G	L	k_0	k_1	k_2	θ	r	degree
Identical or monozygous twin	D/E	0	NA	0	0	1	1/2	1	0
Parent/Offspring	A/F	1	1	0	1	0	$1/2^2$	1/2	1
Sibling	E/F	2	2	1/4	1/2	1/4			
Grand-parent /grand-child	A/I	2	1						
Half-sibling	L/M	2	1	1/2	1/2	0	$1/2^3$	$1/2^2$	2
Avuncular	I/F	3	2						
Great-grandparent	A/P	3	1						
First cousin	I/L	4	2	3/4	1/4	0	$1/2^4$	$1/2^3$	3
Grand avuncular	F/O	4	2						
Great-great-grandparent	-	4	1						
Half-cousin	P/Q	5	2	7/8	1/8	0	$1/2^5$	$1/2^4$	4
Great-grand-avuncular	-	5	2						
Great-great-great grandparent	-	5	1	15/16	1/16	0	$1/2^6$	$1/2^5$	5
Second-cousin	O/P	6	2						
Half-second-cousin	-	7	2	31/32	1/32	0	$1/2^7$	$1/2^6$	6
Third cousin	-	8	2	63/64	1/64	0	$1/2^8$	$1/2^7$	7
Unrelated	-	-	-	1	0	0	0	0	unrelated

on the probability that a random allele from one individual is IBD to the random allele of another makes it impossible to distinguish between a number of possible cases within the same degree. For example, the θ coefficient between sibling pairs and parents and their offspring is equal to 50%. However, the pattern of allele sharing is different. In this case, characterizing the set of IBD sharing coefficients (k_0, k_1 and k_2) is necessary to solve the true degree of relatedness (Table 1). It is noteworthy that even though k_0, k_1 and k_2 coefficients specify the expected fraction of genome sharing between two relatives, the actual sharing measured might strongly differ from these expected values as a result of the random nature of recombination. For instance, the expected IBD sharing between siblings is 50%, while the 95% confidence interval lies between 37 and 63% (Speed and Balding 2015), with variance around the expected 50% value being greater for sibling pairs than for parent-offspring. Other deviations may be introduced by the choice of sequencing technology or relatedness estimators.

Using the power of genomics to infer kinship between ancient individuals

It is now possible to measure genome sharing directly from genome-wide single-nucleotide polymorphism data (SNP), obtained either from HTS DNA sequencing of shotgun or captured libraries (or directly from SNP microarray genotyping when sufficient amounts of fresh DNA are available). To estimate the frequencies of IBD sharing patterns, different approaches have been developed (Table 2). They are either based on the method of moments (Population-based LINKage analyses, PLINK, Purcell et al. 2007), which has the computational efficiency to exploit the potential of large data sets, or on maximum likelihood estimations (Kinship-base INference for Genome-wide association studies, KING, Manichaikul et al. 2010).

In these methods, genetic relatedness is quantified through the calculation of allele-sharing coefficients (Manichaikul et al. 2010), which rely on identifying putative IBD regions from expanding seeds of matching genotypes between individuals (GERMLINE, Genetic Error-tolerant Regional Matching with LInear-time Extension, Gusev et al. 2009). Alternatively, genetic relatedness is calculated on the basis of pre-defined IBD segments (ERSA, Estimation of Recent Shared Ancestry, Huff et al. 2011). These methods can be highly accurate in detecting distant relatedness up to the 13th degree (ERSA, Huff et al. 2011), but their quality requirements make them difficult, if not impossible, to apply to ancient DNA data.

Indeed, for the very vast majority of archaeological samples, no pedigrees are known, and no reference panels providing reliable past allele frequencies in the population are available. In addition, only low-coverage data are generally accessible, and include significant amounts of errors due to post-mortem damage and, in some cases, contamination introduced by one or more present-day individuals. Furthermore, many algorithms assume that the individuals are from the same homogeneous population, without substructure or admixture, and show no significant level of inbreeding. The nature of ancient DNA data, thus, requires the implementation of dedicated statistical methods. These methodologies are developed further below, and account for the specificities of ancient genome-wide data, namely their low coverage, the absence of a contemporaneous reference panel, and the potential occurrence of admixture, population structure, or inbreeding.

Dealing with sparse DNA sequencing data

DNA extracts obtained from ancient remains consist of a mixture of endogenous molecules from the individual of interest, and contaminant molecules deriving from the environment, the laboratory reagents and the scientists handling the remains (e.g., archaeologists and/or the geneticists). The discovery that ancient DNA preservation was maximized in petrous bones or tooth cementum, and the development of molecular

Table 2.
The different relatedness inference methods reviewed:
L: can be applied to **L**ow coverage data;
S: accounts for population **S**tructure;
A: allows inference on **A**dmixed individuals;
I: allows inference on **I**nbred individuals;
P: can be applied even in the absence of a reference **P**anel.

Reference	Software	L	S	A	I	P	Applied/tested on ancient DNA in/from
Purcell et al. 2007	PLINK	-	-	-	-	-	-
Gusev et al. 2009	GERMLINE	-	-	-	-	-	-
Manichaikul et al. 2010	KING-robust	Y	-	-		Y	Sikora et al. 2017 Schroeder et al. 2019
Wang 2011a	COANCESTRY				Y		-
Huff et al. 2011	ERSA	-	-	-	-	-	-
Li et al. 2014	GRAB	-	-	-	-	-	-
Lipatov et al. 2015	lcMLkin	Y	-	-	-	-	Amorim et al. 2018 Daly et al. 2018 Damgaard et al. 2018 Krause-Kyora et al. 2018 Chylenski et al. 2019 Mittnik et al. 2019 Wang et al. 2019 Furtwängler et al. 2020
Korneliussen and Moltke 2015	ngsRelate	Y	-	-	-	-	O'Sullivan et al. 2018 Kuhn et al. 2018
Dou et al. 2017	SEEKIN	Y	Y	Y	-	-	-
Martin et al. 2017	GRUPS	Y	-	-	-	Y	-
Theunert et al. 2017	relcoas	Y	-	-	-	Y	Haak et al. 2015
Waples et al. 2019	IBSrelate	Y	-	-	-	Y	Schroeder et al. 2019
Kuhn et al. 2018	READ	Y	-	-	-	Y	Mathieson et al. 2015 Harney et al. 2018 O'Sullivan et al. 2018 Chylenski et al. 2019 Kashuba et al. 2019 Mittnik et al. 2019 Saag et al. 2019 Sanchez-Quinto et al. 2019 Santana et al. 2019 Scheib et al. 2019 Villalba Mouco et al. 2019 Wang et al. 2019 Furtwängler et al. 2020
Hanghoej et al. 2019	ngsRelateV2	Y	-	-	Y	-	Damgaard et al. 2018

approaches tailor-made to the biochemical nature of ancient molecules have been instrumental in improving the quality of ancient DNA datasets (Orlando et al. 2015). Nevertheless, high coverage genomic data can rarely be obtained at a reasonable cost, and the vast majority of ancient human genomes are limited to low coverage data.

Unfortunately, computational methods adapted to vast modern cohorts, such as PLINK (Purcell et al. 2007), GERMLINE (Gusev et al. 2009) and KING (Manichaikul et al. 2010), require inputs in the form of high-quality genotype data (Fig. 2) and can therefore not be directly applied to the low-depth HTS data obtained from the majority of ancient human remains. Indeed, when the genome is sequenced to low-depth (<5X coverage), there is a high probability that only one of the two alleles has been sampled at a specific site, making accurate genotype calling difficult, if not impossible. Such sparse data may lead to an underestimation of the true level of heterozygosity (Renaud et al. 2018). Therefore, when applied to called genotypes from low coverage data, software tools like PLINK (Purcell et al. 2007) tend to overestimate the fraction of IBD=0 (k_0) and as a consequence underestimate relatedness.

To mitigate the uncertainty when calling genotypes from low-depth HTS data, dedicated programs which incorporate uncertainty of the genotypes in maximum likelihood estimates of pairwise relatedness have been developed (NgsRelate, Korneliussen 2015; lcMLkin, Lipatov et al. 2015). These methods are not based on genotypes, but on genotype likelihoods instead, i.e., on the probability of the observed data in the sequencing read given the true genotype. They thereby take into account the uncertainty of the genotype, described by a quality score or a genotype likelihood that incorporates errors introduced, for instance, by post-mortem damage during sequencing or while mapping to the reference

Fig. 2.
High and Low Coverage sequencing data. When a genomic position is covered by a high number of reads, a high-quality genotype can be called (left). Whereas when only low-coverage data are available, one can benefit from performing analyses that allow uncertainty regarding genotypes and incorporate genotype likelihoods (GL), taking into account errors introduced at different steps (right).

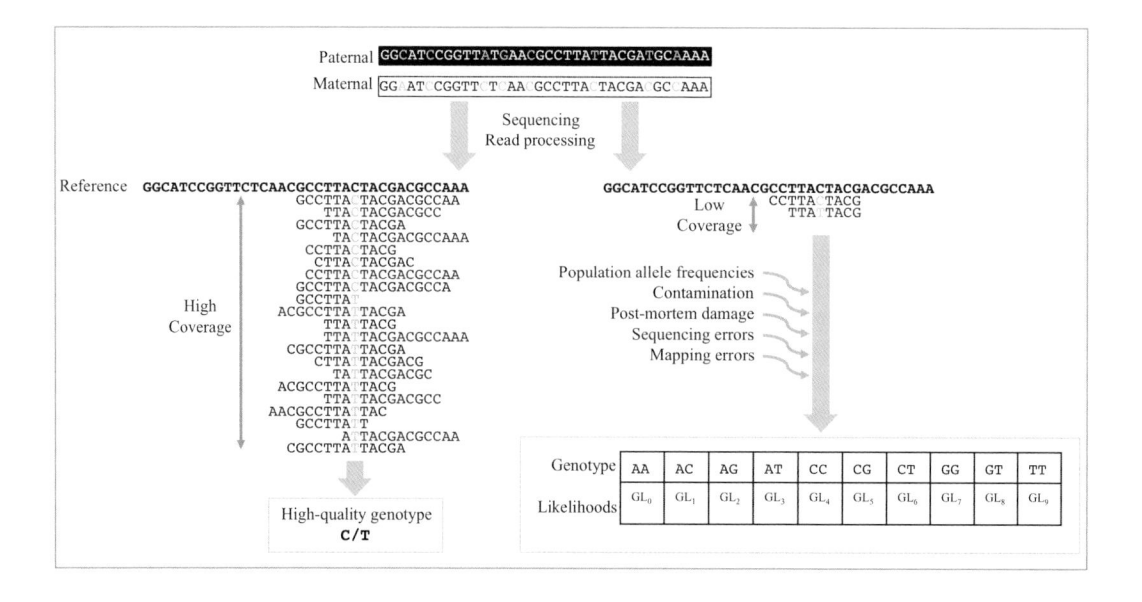

genome (Fig. 2). Of note, these algorithms require similar levels of uncertainty across samples. Applying them to different batches of samples showing different qualities, for instance samples sequenced using different platforms, can bias relatedness estimates. Moreover, they can only be applied to medium-to-low coverage data (2-4X) and not to extremely-low coverage data, unlike other methods based on genetic distances (Martin et al. 2017).

SNP-based methods also require a set of shared SNPs to be covered in all the genomes compared, which can be limiting in low-coverage sequencing contexts. To overcome this, SEEKIN (SEquence-based Estimation of KINship; Dou et al. 2017) leverages the correlation between neighboring markers. Even if the number of common SNPs sequenced in two individuals assessed for relatedness is low, Linkage Disequilibrium (LD) neighboring SNPs can be exploited to call genotypes over SNPs that were not directly sequenced. This results in a significant increase in the total number of SNP positions shared.

One example of successful characterization of family links from extremely low depth sequencing is the identification, 100 years after his death, of the remains of an Irish rebel buried in a shallow grave, which was made possible by the comparison with genetic data from living relatives (Fernandes et al. 2017). The depth of coverage obtained after shotgun sequencing was as low as 0.04X for the deceased, and 0.08-0.1X for presumed second-degree relatives. Since very few SNPs were covered by more than one sequencing read, a forced homozygous approach served as the basis for estimating kinship coefficients (Queller and Goodnight 1989) using the European allele frequencies estimated by "The 1000 Genomes Project" (The 1000 Genomes Project Consortium 2015). Such an approach is efficient in terms of sequencing budget, computing time and accuracy, but can be applicable only in the few cases in which several known close relatives can be included, and when the ancient individual is recent enough that a modern reference panel can be used to assess allelic frequencies. How to proceed when this is not the case, in other words when no proper reference panel can be used to obtain reliable allele frequencies in the population, is the subject of the following chapter section.

How to proceed when no reference panels are available?

Many methods rely on the availability of accurate genotype frequencies from the focal population. Present-day populations can be used as a reference panel only for recent historical individuals (Fernandes et al. 2017). Applying this strategy to more ancient individuals can result in highly-overestimated genetic relatedness, e.g., supporting unrealistic values for individuals living several thousands of years apart (Sikora et al. 2017).

This limitation is not specific to genome-wide SNPs but also applies to STR-based estimations in forensics, where the population to which a suspect belongs is most often unknown (Caliebe and Krawczak 2018), or in genetics applied to historical or archaeological samples (Cui et al. 2015; Russo et al. 2016). For instance, funerary practices in the territory of Argentina before Hispanic contact was estimated based on previously published STR data from other pre-Hispanic populations from Peru (Baca et al. 2012), but also, as these data were not sufficient to serve as a reference panel on their own, from contemporary populations (Russo et al. 2016).

Methods that do not necessitate a reference panel still require at least a pair of unrelated individuals from the same population for normalization (READ, Relationship Estimation from Ancient DNA, Kuhn et al. 2018), or genomic data from several other ancient individuals of the same population (Martin et al. 2017; Theunert et al. 2017). The smaller the number of individuals available, the less accurate the kinship coefficient, regardless of the number of genomic sites considered. For example, simulation studies show that with as few as 18 individuals, the method developed by Theunert and colleagues cannot identify second-degree relatives (Theunert et al. 2017). In addition, approaches like READ (Kuhn et al. 2018) are sensitive to batch effects (shotgun versus capture, or data obtained with different sequencing platforms), which can overestimate the genetic distance indicative of unrelated individuals and, thus, the degree of relatedness in the tested samples.

Without access to data from an unrelated pair of the same population, only very few methods can be applied to very small datasets, for comparing a single pair of individuals or to establish relatedness within a single pedigree. KING, for example, is a robust algorithm that can infer relatedness up to the third degree based on information from only two individuals analyzed, but only when high-density genotyping data are available (Manichaikul et al. 2010).

How to proceed in the presence of admixture and population structure?

As mentioned earlier, most algorithms used to estimate relatedness assume that the individuals analyzed belong to a genetically homogeneous population (Purcell et al. 2007; Milligan 2003; Albrechtsen et al. 2009). In the case of admixture, the individuals show different ancestry backgrounds, which violates this assumption. In the case of population structure, the relatedness between individuals belonging to the same sub-population will be systematically inflated. As a consequence, many approaches face circularity: while the identification of unrelated individuals is a prerequisite to the detection of population structure (Zhu et al. 2008), proper relatedness inference relies on the identification of homogeneous populations (Purcell et al. 2007).

Several methods have been developed to overcome this limitation and provide kinship estimation for admixed individuals (Wang 2011b; Relatedness Estimation in Admixed Populations, REAP, Thornton et al. 2012; and RelateAdmix, Moltke and Albrechtsen 2014, also providing admixture proportions). To account for the different ancestry backgrounds in admixed individuals, these approaches use allele frequencies specific to each individual derived from software like ADMIXTURE (Alexander et al. 2009). The KING-robust estimator (Manichaikul et al. 2010) is designed to deal with population structure, but shows lower performance when analyzing admixed individuals (Thornton et al. 2012). However, the estimators mentioned above require accurate genotype data, which may not be available in the majority of paleogenomic studies thus far. The SEEKIN estimator (Dou et al. 2017) can infer kinship for both homogeneous and heterogeneous samples with population structure and admixture. It was initially designed to improve kinship estimation in target-enrichment experiments by incorporating off-target reads (0.15X average depth), but can be advantageously applied to shallow sequencing data.

Inbreeding

All the relatedness estimators reviewed so far in this chapter indicate that the individuals investigated are not inbred. However, in archaeological contexts, suspicions that this is not the case are worth considering. For example, past hunter-gatherers were previously found to present a high level of skeletal abnormalities, which was interpreted as the consequence of inbreeding among small human groups (Trinkaus et al. 2014). High inbreeding levels were not, however, confirmed by ancient DNA data (Sikora et al. 2017). Can small amounts of inbreeding affect kinship estimates? How shall we treat data from dynasties with a known history of consanguinity?

At the genomic level, inbreeding results in an excess of homozygosity compared to what would be expected under Hardy-Weinberg Equilibrium, which describes the relationship between allele frequencies and genotype frequencies in a randomly mating population (see Andrews 2010 for a review of the basics). In this case, estimating only k_0, k_1 and k_2 is insufficient, and the nine condensed Jacquard coefficients have to be determined (see section 'How should relatedness degrees and coefficients be understood'). Some methods like COANCESTRY (Wang 2011a) have been developed to estimate relatedness in the presence of inbreeding, but they cannot be applied to low coverage data as they are based on high quality genotypes. At present, ngsRelate v2 (Hanghøj et al. 2019) is the only software based on genotype likelihoods that can infer relatedness in the presence of inbreeding while providing inbreeding coefficients for both individuals considered.

Paleogenomicists now have at hand a whole suite of software enabling them to exploit the power of genomic data in order to assess how much two past individuals were in fact related from a biological point of view. In the following, I present some of the key studies leveraging these tools to facilitate sample curation and for helping solve long-standing archaeological disputes.

PART II: APPLICATIONS

This section will review the major recent publications making use of the statistical methods for kinship inferences between ancient individuals that we detailed in Part I. It will cover a wide range of applications, from data curation by collapsing sample duplicates or removing related individuals from the dataset prior to population genetics studies, to questioning the definition of a 'family' in the archaeological past and elucidating some aspects of social structure in past societies.

Relatedness inferences as a tool for sample and data curation

One of the very first and most straightforward applications of relatedness inference on ancient remains was to reveal unidentified duplicated samples (Table 3). The fragmentation and dispersion of bones found in collective burials and domestic waste, or of remains stored in bulk for many years in museum collections, make it difficult to certify that each and every fragment analyzed is from a different individual. Restricting the analyses to petrous bones from the same side of the skull (Daly et al. 2018) or to teeth in connection to the mandible can limit spurious re-sampling, but is not always an option. After shallow shotgun sequencing or targeted enrichment, systematic processing of relatedness estimation between pairs of samples has revealed kinship coefficients indicative of monozygous twins, most likely representing independent remains from the same individual (Daly et al. 2018; Villalba-Mouco et al. 2019; Wang et al. 2019; Theunert et al. 2017 reanalyzing data from Haak et al. 2015). In the majority of cases, data from both samples can simply be collapsed into one single individual, but in a few others, they should be discarded (Daly et al. 2018 data coming from petrous bones of the same side). Moreover, testing for potential relatedness may be useful if artifacts were produced by the same individual. Indeed, birch bark mastics, commonly used from the Middle Paleolithic onwards as chewing gum but also as an adhesive in lithic tool technology, can fossilize teeth or finger imprints and very surprisingly were also identified as ancient DNA reservoirs, opening the possibility of a direct connection between genetics and a material culture (Kashuba et al. 2019). For example, low-coverage genomes (0.1 to 0.5X) could be reconstructed from three of these pre-historic mastics, excavated in the Mesolithic site of Huseby Klev, Sweden (Kashuba et al. 2019). By performing kinship analyses between these samples, the authors could confirm each of them was processed by a different individual, but the low coverages obtained, combined with the very limited number of samples and the absence of a reference panel, prevented any conclusions to be drawn on their relatedness.

Kinship coefficients can also be used to test for the presence of potential cross-contamination between samples processed during the same experimental sessions, which would bear relatedness across different sites and/or time periods (Villalba-Mouco et al. 2019). Identifying first or second-degree parents in an ancient panel is also crucial, as including those may bias population frequencies-based statistics. When present, relatives with the lower coverage are filtered out of assessments on population genetics (Damgaard et al. 2018; Daly et al. 2018; Harney et al. 2018; Chylenski et al. 2019; Wang et al. 2019). Similarly, testing for the absence of related individuals is necessary before validating their inclusion into a panel for identifying a susceptibility locus to certain diseases (e.g., leprosy, Krause-Kyora et al. 2018).

Table 3.
Overview of the ancient DNA studies including relatedness inference analysis.

Reference	Method used	Type of sequencing	Application
Kennett et al. 2017	Kennett et al. 2017	1240k capture	Kinship
Fernandes et al. 2017	Hardy & Vekemans 2002	shotgun	Remains identification
Sikora et al. 2017	Manichaikul et al. 2010	shotgun	Kinship
Amorim et al. 2018	Lipatov et al. 2015	shotgun + 1240K capture	Kinship
Daly et al. 2018	Lipatov et al. 2015	shotgun	Exclusion of related individuals and collapse of duplicated samples
Damgaard et al. 2018	Lipatov et al. 2015	shotgun	Exclusion of related individuals
Harney et al. 2018	Kuhn et al. 2018	1240k capture	Exclusion of related individuals
Krause-Kyora et al. 2018	Lipatov et al. 2015	shotgun	Exclusion of related individuals
O'Sullivan et al. 2018	Kuhn et al. 2018 Kennett et al. 2017	1240k capture	Kinship
Chylenski et al. 2019	Kuhn et al. 2018 Lipatov et al. 2015	shotgun	Kinship and exclusion of related individuals
Kashuba et al. 2019	Kuhn et al. 2018	shotgun	Check for identical individuals
Mittnik et al. 2019	Kuhn et al. 2018 Lipatov et al. 2015	1240k capture	Exclusion of related individuals Kinship
Sanchez-Quinto et al. 2019	Kuhn et al. 2018	Whole genome capture	Kinship
Santana et al. 2019	Kuhn et al. 2018	Illumina MEGA capture	Kinship
Saag et al. 2019	Kuhn et al. 2018	shotgun	Kinship
Schroeder et al. 2019	Waples et al. 2019 Manichaikul et al. 2010	shotgun	Kinship
Scheib et al. 2019	Kuhn et al. 2018 Chang et al. 2015	shotgun	Kinship
Villalba-Mouco et al. 2019	Kuhn et al. 2018	1240k capture	Collapse of duplicated samples
Wang et al. 2019	Kuhn et al. 2018 Lipatov et al. 2015	1240k capture	Exclusion of related individuals and collapse of duplicated samples
Furtwängler et al. 2020	Kuhn et al. 2018 Lipatov et al. 2015	1240k capture	Kinship

The emergence of the nuclear family as the basis of social relationships

In present-day societies, biological relatedness is the foundation for many of our social interactions, including heritage, marriage rules, parental authority, support obligation, etc. When did this start? Was genetic kinship such a key driver of social organization in the past?

Determining the genetic kinship of past individuals buried together or in a close proximity provides important insights into the social organization in (pre-)historic cultures. To infer potential familial links, and ultimately reconstruct pedigrees or lineages, measures of relatedness obtained from genomic data provides one good starting point. It should, however, always be combined with stratigraphic information, radiocarbon dates, the age and the sex of the individual, etc. This multi-proxy approach was applied on the individuals buried at Sunghir (Russia), which is dated to ~35,000 years BP, and hosts two of the most extraordinary Upper Paleolithic burials known to date (Sikora et al. 2017). In a first pit, an adult male, Sunghir 1, was buried together with thousands of beads made of mammoth ivory, most likely sewn on his clothes. In the other burial, two juvenile individuals, Sunghir 2 and Sunghir 3, were interred head-to-head. They were also both covered in approximately 10,000 beads, but these were slightly smaller as if they were scaled to the height of the children. Among the breath-taking grave goods excavated, we can cite a belt made of almost 300 pierced fox canines, as well as a dozen ivory spears, including one that was 2.5 m long. Surprisingly, a human femoral diaphysis from a fourth individual was found next to Sunghir 2; it was broken, polished and filled with ochre. The Sunghir site represents one of the exceptional cases of multiple individuals buried simultaneously, or originating from very close temporal and spatial proximity. These individuals may thus represent a single social group and provide unprecedented information on the behavior and kinship structure in an Upper Paleolithic society. Based on the artifacts and anatomical observations, the three individuals were often interpreted as members of the same nuclear family: Sunghir 1 as the father of a son (Sunghir 2) and a daughter (Sunghir 3) (Trinkhaus et al. 2014). Genome analyses told a different story, first revealing all four individuals as males (Sikora et al. 2017). All four individuals also carried the same mitochondrial genome sub-haplogroup U2 (in accordance with a West Eurasian and Siberian Paleolithic background) and the same Y-chromosome haplogroup C1a2. Despite this, and the anatomical similarities observed, none of the Sunghir individuals were found to be closely related, at least not up to the third degree, as indicated by a method that does not rely on unknown allele frequencies for Upper Paleolithic populations (Manichaikul et al. 2010).

Inbreeding was also examined for Sunghir 3, as the skeletal pathologies that his remains displayed were previously interpreted as evidence for elevated inbreeding. The cumulative length of long Run Of Homozygosity (ROH), used as a proxy for inbreeding, was found to be higher than in most ancient genomes, but shorter than in archaic and modern populations with a known history of isolation and consanguinity (Sikora et al. 2017). This rules out recent inbreeding in Sunghir individuals despite small population size, which indicates that groups of restricted size and limited kinship were embedded in a larger mating network, similar to what is observed in present-day hunter-gatherers. Even though the debate about active and conscious avoidance of consanguinity is still open, the Sunghir genomes support the existence of regular exchanges of mates between groups, which is also consistent with archaeological evidence of high mobility in the Upper Paleolithic, and may have impacted the development of knowledge transfer between bands.

The earliest molecular identification of a nuclear family (mother-father-child) dates back to ~ 30,000 years later. Schroeder and colleagues (Schroeder et al. 2019) have performed a systematic analysis of the genomes of 15 late Neolithic individuals buried in a mass grave at Koszyce, in today's Southern Poland (2,880-2,776 BCE). As in Sikora et al. 2017, the authors used a method based on pairwise sharing of alleles IBS (see section 'From pedigree to relatedness inference, and back'), similar to the one formalized by Waples and colleagues (Waples et al. 2019) and combined with kinship estimators (KING, Manichaikul et al. 2010). This combination allowed the authors not only to discriminate among the different first three degrees of relatedness, but also, within first degree related pairs, to identify parent-offspring or full-siblings (Table 1). Relatedness inferences showed that individuals were positioned in the grave according to kindred relations, highlighting that social and biological relationships could, at least to some extent, be superimposed, and confirming that family relations were key to the community organization. In particular, according to the arrangement of the deceased within the pit, maternal kinship and brotherhood were considered as an important form of social investment, important enough to be perpetuated in death. This mass murder situation is also an exceptional case where several individuals were buried simultaneously and immediately after their death, making the hypothesis of corpse repatriation very unlikely: obviously, people were buried in the exact place where they lived just before their death. Of note, even though results are consistent with the community being organized along patrilineal lines of descent, women in different family positions (mothers, daughters, sisters) are present in the burial. Interestingly, both daughters identified were sub-adults (13-14 and 15-16 years old), whereas four of the seven identified sons were already adults (and a fifth at 16-17 years old). Therefore, these observations are consistent with the hypothesis of patrilocality, where women leave their parents and find a partner in another location while men stay, or at least come back as adults, to the community where they were born.

In a more recent study, including 13 sites dated from the Late Neolithic to the Bronze Age and located in Switzerland, South Germany and French Alsace, 27 of 96 individuals whose genome-wide data have been analyzed were identified using lcMLkin (Lipatov et al. 2015) and READ (Kuhn et al. 2018) to be related at the first or second degree, within nine familial groups (Furtwängler et al. 2020). This result underlies the importance of kinship in these societies. Strikingly, the vast majority of these related individuals were males (22/27) and all but one of the females were mothers of one or two identified sons, adding some arguments to the hypothesis of patrilocality and social importance of the male lineage.

Both case studies above (Schroeder et al. 2019; Furtwängler et al. 2020) identified full brother/sisterhoods are largely predominant in numbers over half-sisterhoods, and the only half-brothers identified have the same father (but a different mother). This observation, which can only be accessed through genetic relatedness inference, is key for understanding the nature of a family nucleus during the Late Neolithic and Bronze Age in Germany and Switzerland, with most likely a stable parental couple between a local male and a non-local female, and possibly polygamy or re-union for men. In the next section, further examples provide information on unilineal descent groups, as well as the correlation between these mating rules and social stratification.

Complex societies and unilineal descent groups

There is a long-standing debate, in anthropology and in the social sciences in general, about the importance of kinship and unilineal (matrilineal or patrilineal) descent groups in modeling the structure, hierarchy and evolution of complex societies. What are the rules underlying the choice of high-ranking members of a given community? Do biological factors, such as kinship or age, dominate social rank? Or is leadership, which is based on life-time achievements such as success at war or hunting, more important? Inferring relatedness among members of a burial that could be defined as an elite burial through its architecture, context, grave goods or personal adornments can help address the hypothesis of potentially institutionalized heredity of leadership in pre-historic societies. Finding a link between the presence of related individuals and rich grave goods can indicate that material wealth or elite status was transferred from parent to offspring.

Pueblo Benito, New Mexico, was both the spiritual and political center of the Chacoan society, one of North America's earliest complex societies (Chaco Canyon, 800 CE-1,130 CE). By analyzing genome-wide capture data from individuals interred in one of its most elaborate elite burials, Kennett and colleagues revealed a community organized as a matrilineal dynasty (Kennett et al. 2017). They calculated relatedness coefficients with an approach similar to the one formalized by Kuhn and colleagues (Kuhn et al. 2018). Here, genomes were pseudohaploidized following random sampling of an allele. The average mismatch rate across all autosomal SNPs was then computed and normalized by the highest mismatch rate observed among all the individuals, assuming that the corresponding specific pair of individuals belonged to the same population and were unrelated. This is the first study of its kind that confirmed the importance of genetic kindred in highly complex and structured societies.

In contrast, studies investigating past European societies revealed social organization based around male lines of descents. For example, this was true for the Lombards, who ruled over part of Italy for over 200 years and took part in the migrations that shaped European societies during the 4th-6th century CE. As written records about these migrations are highly partial and settlements barely known, questions about their identity and social organization could only be addressed through funerary archaeology and ancient DNA. Two Lombard cemeteries, located in Hungary and Northern Italy, were sampled and screened for DNA preservation (Amorim et al. 2018). A total of 55 individuals provided sufficient genome-wide capture data to be scanned for relatedness using lcMLkin (Lipatov et al. 2015). As reference allele frequencies, the authors used those of the ancient individuals themselves combined with those from the present-day 1000 Genome populations. In addition, they adapted the lcMLkin software to incorporate admixture and to account for diverse genetic ancestry. This study highlighted the fact that groups interred in both cemeteries appeared to be organized around one extended male kindred of high status, according to their level of meat consumption and the associated artifacts and grave goods. This conclusion underlined that biological relatedness may have played a major role in the structure and hierarchy of these societies.

A recent study on individuals excavated from a 7th century CE Alemanni site in southern Germany reminds us that social structure involves more subtle factors than single, linear familial transmission of culture and power. Here, the burials delivered extremely rich grave goods, including jewelry, equestrian gear, and weapons, demonstrating the wealth and power of the household, and suggesting contacts with Byzantines, Lombards and Franks. Genome-wide SNP enrichment could be performed on a selection of eight male individuals (O'Sullivan et al. 2018) and kinship was estimated based on the proportion of non-matching autosomal genotypes (Kennett et al. 2017; Kuhn et al. 2018). Five individuals, coming from five different graves, including two multiple burials, and accompanied by culturally different artifacts, appeared as first and second-degree relatives. This surprising result underlines that burial patterns and assignment of grave goods do not necessarily reflect genetic relatedness. It also highlights that cultural appropriation from diverse origins can be found even among relatives as close as father and son. On the other hand, the presence of non-relatives within the same grave suggests that social fellowship could be held equally as with biological relatedness in this Alemanni funerary site. This confirms analyses of historical records indicating that, in the Merovingian period, being part of a household was not limited to biological relatives.

As described earlier, such patrilineal social organization in western Europe seems to extend back at least to the late Neolithic (Schroeder et al. 2019; Furtwängler et al. 2020). Mittnik and colleagues have recently performed an archaeo-genomic study integrating genome-wide capture data, material culture, as well as strontium and oxygen stable isotope analyses of 104 individuals from the Lech Valley, Germany, spanning from the Late Neolithic to the Middle Bronze Age (Mittnik et al. 2019). They discuss the hypothesis that households were, at that time and region, socially stratified institutions. Indeed, in the vicinity of farmsteads, three different types of burials were identified: those associated with rich grave goods and holding the remains of members of the same family; those well-furnished but where non-local adult women were buried; and a third burial type with low-status, unrelated women. Interestingly, among the 10 different parent-offspring relations identified, the offspring was systematically a male (adult in 9 out of 10 cases), indicating that daughters may have had to leave their parental home. This observation is consistent with isotopic and archaeological evidence suggesting that high-status women identified on site were from non-local origin, most likely coming from several hundreds of kilometers away. Of note, these studies (Schroeder et al. 2019; Mittnik et al. 2019; Furtwängler et al. 2020) focus only on Central Europe (Poland, Germany, Switzerland). Considerable extra sampling and sequencing efforts are still needed before their conclusions can be used to generalize on a wider European scale.

CONCLUSION

A wide range of methods and software tools have been designed for inferring relatedness from genomic or genome-wide SNP data. A total of 12 methods have been benchmarked, showing high accuracy for first- and second-degree relationships in modern DNA data (Ramstetter et al. 2017). IBD-based methods, however, appeared more accurate and more efficient in deciphering distant relatedness than methods relying on independent markers. The idiosyncrasies of ancient DNA data, such as the absence of reference panels, the diverse molecular tools or platforms used, the uncertainty on strict sample contemporaneity, or the various coverage achieved across individuals, preclude similar benchmarking on real data. The power and accuracy of relatedness inference methods can, however, be evaluated through simulations (Martin et al. 2017; Hanghøj et al. 2019) or on small sample size datasets (Theunert et al. 2017). Methods developed for shallow genome sequencing can not only benefit forensics and ancient DNA studies, but also be applied to wild animal survey through non-invasive sample collections (e.g., fecal baboon DNA, Snyder-Mackler et al. 2016), off-target regions in target sequencing studies (Dou et al. 2017), or very large sample size studies where only very low coverage sequencing is affordable. The best strategy remains debatable, especially as combining the inferences from different methods does not seem to drastically enhance the accuracy of the relatedness inferences drawn (Ramstetter et al. 2017). Future approaches may not be restricted to pairwise comparisons, but instead will also exploit the relatedness signals obtained from multiple individuals.

Applied to ancient humans, these molecular analyses aim at serving as tools to complement historical and archaeological approaches in order to get a deeper understanding of past societies. For many years, major paleogenomic studies have been based on sampling individuals across multiple sites, assumed to be representative of a defined culture (Haak et al. 2015; Allentoft et al. 2015; Olalde et al. 2018). Some of the most recent studies favor a micro-regional approach and attempt to characterize the genomes of all possible individuals in a multiple/collective burial or in a cemetery, aiming at revealing its entire genomic complexity. In combination with the study of material culture, funerary practices, isotopic data, physical anthropology, paleopathology and others, molecular approaches to kinship inference will help us better understand the rules and factors underpinning the organization of past societies.

ACKNOWLEDGMENTS

The author is grateful to Clio Der Sarkissian, Antoine Fages and Ludovic Orlando for their critical reading of the manuscript. This project has received funding from the European Union's Horizon 2020 research and innovation programme under the Marie Sklodowska-Curie grant agreement No 795916.

REFERENCES

Albrechtsen, A., T. S. Korneliussen, I. Moltke, T. van Overseem Hansen, F. C. Nielsen, and R. Nielsen. 2009. Relatedness mapping and tracts of relatedness for genome-wide data in the presence of linkage disequilibrium. *Genetic Epidemiology* 33: 266–274.

Alexander, D. H., J. Novembre, and K. Lange. 2009. Fast model-based estimation of ancestry in unrelated individuals. *Genome Research* 19: 1655–1664.

Allentoft, M. E., M. Sikora, K.-G. Sjogren, S. Rasmussen, M. Rasmussen, J. Stenderup, P. B. Damgaard, H. Schroeder, T. Ahlstrom, L. Vinner, A.-S. Malaspinas, A. Margaryan, T. Higham, D. Chivall, N. Lynnerup, L. Harvig, J. Baron, P. Della Casa, P. Dabrowski, P. R. Duffy, A. V. Ebel, A. Epimakhov, K. Frei, M. Furmanek, T. Gralak, A. Gromov, S. Gronkiewicz, G. Grupe, T. Hajdu, R. Jarysz, V. Khartanovich, A. Khokhlov, V. Kiss, J. Kolar, A. Kriiska, I. Lasak, C. Longhi, G. McGlynn, A. Merkevicius, I. Merkyte, M. Metspalu, R. Mkrtchyan, V. Moiseyev, L. Paja, G. Palfi, D. Pokutta, Ł Pospieszny, T. D. Price, L. Saag, M. Sablin, N. Shishlina, V. Smrcka, V. I. Soenov, V Szeverenyi, G Toth, S. V. Trifanova, L. Varul, M. Vicze, L. Yepiskoposyan, V. Zhitenev, L. Orlando, T. Sicheritz-Ponten, S. Brunak, R. Nielsen, K. Kristiansen and E. Willerslev. 2015. Population genomics of Bronze Age Eurasia. *Nature* 522: 167–172.

Amorim, C. E. G., S. Vai, C. Posth, A. Modi, I. Koncz, S. Hakenbeck, M. C. La Rocca, B. Mende, D. Bobo, W. Pohl, L. Pejrani Baricco, E. Bedini, P. Francalacci, C. Giostra, T. Vida, D. Winger, U. von Freeden, S. Ghirotto, M. Lari, G. Barbujani, J. Krause, D. Caramelli, P. J. Geary, and K. R. Veeramah. 2018. Understanding 6th-century barbarian social organization and migration through paleogenomics. *Nature Communications* 9: 3547.

Andrews, C. 2010. The Hardy-Weinberg Principle. *Nature Education Knowledge* 3 (10): 65.

Baca, M., K. Doan, M. Sobczyk, A. Stankovic, and P. Węgleński. 2012. Ancient DNA reveals kinship burial patterns of a pre-Columbian Andean community. *BMC Genetics* 13: 30.

Brenner, C. H. and B.S. Weir. 2003. Issues and strategies in the DNA identification of World Trade Center victims. *Theoretical Population Biology* 63: 173–178.

Caliebe, A., and M. Krawczak. 2018. Match probabilities for Y-chromosomal profiles: A paradigm shift. *Forensic Science International: Genetics* 37: 200–203.

Chang, C. C., C. C. Chow, L. C. Tellier, S. Vattikuti, S. M. Purcell, and J. J. Lee. 2015. Second-generation PLINK: Rising to the challenge of larger and richer datasets. *Gigascience* 4: 7.

Chylenski, M., E. Ehler, M. Somel, R. Yaka, M. Krzewinska, M. Dabert, A. Juras, and A. Marciniak. 2019. Ancient mitochondrial genomes reveal the absence of maternal kinship in the burials of Çatalhöyük people and their genetic affinities. *Genes* 10: 3390.

Cui, Y., L. Song, D. Wei, Y. Pang, N. Wang, C. Ning, C. Li, B. Feng, W. Tang, H. Li, Y. Ren, C. Zhang, Y. Huang, Y. Hu, and H. Zhou. 2015. Identification of kinship and occupant status in Mongolian noble burials of the Yuan Dynasty through a multidisciplinary approach. *Philosophical Transactions of the Royal Society* B 370: 20130378.

Daly, K. G., P. M. Delser, V. E. Mullin, A. Scheu, V. Mattiangeli, M. D. Teasdale, A. J. Hare, J. Burger, M. P. Verdugo, M. J. Collins, R. Kehati, C. M. Erek, G. Bar-Oz, F. Pompanon, T. Cumer, C. Çakırlar, A. F. Mohaseb, D. Decruyenaere, H. Davoudi, Ö. Çevik, G. Rollefson, J.-D. Vigne, R. Khazaeli, H. Fathi, S. B. Doost, R. R. Sorkhani, A. A. Vahdati, E. W. Sauer, H. A. Kharanaghi, S. Maziar, B. Gasparian, R. Pinhasi, L. Martin, D. Orton, B. S. Arbuckle, N. Benecke, A. Manica, L. K. Horwitz, M. Mashkour, and D. G. Bradley. 2018. Ancient goat genomes reveal mosaic domestication in the Fertile Crescent. *Science* 361: 85–88

Damgaard, P. d. B., R. Martiniano, J. Kamm, J. V. Moreno-Mayar, G. Kroonen, M. Peyrot, G. Barjamovic, S. Rasmussen, C. Zacho, N. Baimukhanov, V. Zaibert, V. Merz, A. Biddanda, I. Merz, V. Loman, V. Evdokimov, E. Usmanova, B. Hemphill, A. Seguin-Orlando, F. E. Yediay, I. Ullah, K.-G. Sjögren, K. H. Iversen, J. Choin, C. de la Fuente, M. Ilardo, H. Schroeder, V. Moiseyev, A. Gromov, A. Polyakov, S. Omura, S. Y. Senyurt, H. Ahmad, C. McKenzie, A. Margaryan, A. Hameed, A. Samad, N. Gul, Muhammad Hassan Khokhar25, O. I. Goriunova, V. I. Bazaliiskii, J. Novembre, A. W. Weber, L. Orlando, M. E. Allentoft, R. Nielsen, K. Kristiansen, M. Sikora, A. K. Outram, R. Durbin, and E. Willerslev. 2018. The first horse herders and the impact of early Bronze Age steppe expansions into Asia. *Science* 360: eaar7711.

Deguilloux, M. F., M. H. Pemonge, F. Mendisco, D. Thibon, I. Cartron, and D. Castex. 2013. Ancient DNA and kinship analysis of human remains deposited in Merovingian necropolis sarcophagi (Jau Dignac et Loirac, France, 7th-8th century AD). *Journal of Archaeological Science* 41: 399–405.

Dou, J., B. Sun, X. Sim, J. D. Hughes, D. F. Reilly, E. S. Tai, J. Liu, and C. Wang. 2017. Estimation of kinship coefficient in structured and admixed populations using sparse sequencing data. *PLoS Genetics* 13: e1007021.

Fernandes, D., K. Sirak, M. Novak, J. A. Finarelli, J. Byrne, E. Connolly, J. E. L. Carlsson, E. Ferretti, R. Pinhasi, and J. Carlsson. 2017. The identification of a 1916 Irish rebel: New approach for estimating relatedness from low coverage homozygous genomes. *Scientific Reports* 7: 41529.

Furtwängler, A., A. B. Rohrlach, T. C. Lamnidis, L. Papac, G. U. Neumann, I. Siebke, E. Reiter, N. Steuri, J. Hald, A. Denaire, B. Schnitzler, J. Wahl, M. Ramstein, V. J. Schuenemann, P. W. Stockhammer, A. Hafner, S. Lösch, W. Haak, S. Schiffels, and J. Krause. 2020. Ancient genomes reveal social and genetic structure of Late Neolithic Switzerland. *Nature Communications* 11: 1915.

Gerstenberger, J., S. Hummel, T. Schultes, B. Hack, and B. Herrmann. 1999. Reconstruction of a historical genealogy by means of STR analysis and Y-haplotyping of ancient DNA. *European Journal of Human Genetics* 7: 469–477.

Gusev, A., J. K. Lowe, M. Stoffel, M. J. Daly, D. Altshuler, J. L. Breslow, J. M. Friedman, and I. Pe'er. 2009. Whole population, genome-wide mapping of hidden relatedness. *Genome Research* 19: 318–326.

Haak, W., G. Brandt, H. N. de Jong, C. Meyer, R. Ganslmeier, V. Heyd, C. Hawkesworth, A. W. G. Pike, H. Meller, and K. W. Alt. 2008. Ancient DNA, Strontium isotopes, and osteological analyses shed light on social and kinship organization of the Later Stone Age. *PNAS* 105: 18226–18231.

Haak, W., I. Lazaridis, N. Patterson, N. Rohland, S. Mallick, B. Llamas, G. Brandt, S. Nordenfelt, E. Harney, K. Stewardson, Q. Fu, A. Mittnik, E. Banffy, C. Economou, M. Francken, S. Friederich, R. Garrido Pena, F. Hallgren, V. Khartanovich, A. Khokhlov, M. Kunst, P. Kuznetsov, H. Meller, O. Mochalov, V. Moiseyev, N. Nicklisch, S. L. Pichler, R. Risch, M. A. Rojo Guerra, C. Roth, A. Szecsenyi-Nagy, J. Wahl, M. Meyer, J. Krause, D. Brown, D. Anthony, A. Cooper, K. Werner Alt, and D. Reich. 2015. Massive migration from the steppe was a source for Indo-European languages in Europe. *Nature* 522: 207–211.

Han, L., and M. Abney. 2011. Identity by descent estimation with dense genome-wide genotype data. *Genetic Epidemiology* 35: 557–567.

Hanghøj, K., I. Moltke, P. A. Andersen, A. Manica and T. S. Korneliussen. 2019. Fast and accurate relatedness estimation from high-throughput sequencing data in the presence of inbreeding. *GigaScience* 8: 1–9.

Hardy, O. J., and X. Vekemans. 2002. Spagedi: A versatile computer program to analyse spatial genetic structure at the individual or population levels. *Molecular Ecology Notes* 2: 618–620.

Harney, E., H. May, D. Shalem, N. Rohland, S. Mallick, I. Lazaridis, R. Sarig, K. Stewardson, S. Nordenfelt, N. Patterson, I. Hershkovitz, and D. Reich. 2018. Ancient DNA from Chalcolithic Israel reveals the role of population mixture in cultural transformation. *Nature Communications* 9: 3336.

Huff, C. D., D. J. Witherspoon, T. S. Simonson, J. Xing, W. S. Watkins, Y. Zhang, T. M. Tuohy, D. W. Neklason, R. W. Burt, S. L. Guthery, S. R. Woodward, and L. B. Jorde. 2011. Maximum-likelihood estimation of recent shared ancestry (ERSA). *Genome Research* 21: 768–774.

Jacquard, A. 1974. *The Genetic Structure of Populations*. Biomathematics. Vol. 5. Berlin: Springer.

Kashuba, N., E. Kırdök, H. Damlien, M. A. Manninen, B. Nordqvist, P. Persson and A. Götherström. 2019. Ancient DNA from mastics solidifies connection between material culture and genetics of mesolithic hunter-gatherers in Scandinavia. *Communications Biology* 2: 185.

Kennett, D. J., S. Plog, R. J. George, B. J. Culleton, A. S. Watson, P. Skoglund, N. Rohland, S. Mallick, K. Stewardson, L. Kistler, S. A. LeBlanc, P. M. Whiteley, D. Reich, and G. H. Perry. 2017. Archaeogenomic evidence reveals prehistoric matrilineal dynasty. *Nature Communications* 8: 14115.

Kennett, D. A., A. Timpson, D. J. Balding, and M. G. Thomas. 2018. The rise and fall of Britains DNA: A tale of misleading claims, media manipulation and threats to academic freedom. *Geneaolgy* 2: 47.

Knipper, C., A. Mittnik, K. Massy, C. Kociumaka, I. Kucukkalipci, M. Maus, F. Wittenborn, S. E. Metz, A. Staskiewicz, J. Krause, and P. W. Stockhammer. 2017. Female exogamy and gene pool diversification at the transition from the Final Neolithic to the Early Bronze Age in central Europe. *PNAS* 114: 10083–10088.

Korneliussen, T. S., and I. Moltke. 2015. NgsRelate: A software tool for estimating pairwise relatedness from next-generation sequencing data. *Bioinformatics* 31: 4009–4011.

Krause-Kyora, B., M. Nutsua, L. Boehme, F. Pierini, D. D. Pedersen, S.-C. Kornell, D. Drichel, M. Bonazzi, L. Möbus, P. Tarp, J. Susat, E. Bosse, B. Willburger, A. H. Schmidt, J. Sauter, A. Franke, M. Wittig, A. Caliebe, M. Nothnagel, S. Schreiber, J. L. Boldsen, T. L. Lenz, and A. Nebel. 2018. Ancient DNA study reveals HLA susceptibility locus for leprosy in medieval Europeans. *Nature Communications* 9: 1569.

Kuhn, J. M. M., M. Jakobsson, and T. Gunther. 2018. Estimating genetic kin relationships in prehistoric populations. *PLoS ONE* 13: e0195491.

Li, H., G. Glusman, C. Huff, J. Caballero, and J. C. Roach. 2014. Accurate and robust prediction of genetic relationship from whole-genome sequences. *PLoS ONE* 9: e85437.

Lipatov, M., K. Sanjeevy, R. Patroy, and K. R. Veeramah. 2015. Maximum likelihood estimation of biological relatedness from low coverage sequencing data. bioRxiv. Cold Spring Harbor Labs Journals doi:10.1101/023374.

Lynch, M., and K. Ritland. 1999. Estimation of pairwise relatedness with molecular markers. *Genetics* 152: 1753–1766.

Manichaikul, A., J. C. Mychaleckyj, S. S. Rich, K. Daly, M. Sale, and W.-M. Chen. 2010. Robust relationship inference in genome-wide association studies. *Bioinformatics* 26: 2867–2873.

Martin, M. D., F. Jay, S. Castellano, and M. Slatkin. 2017. Determination of genetic relatedness from low-coverage human genome sequences using pedigree simulations. *Molecular Ecology* 26: 4145–4157.

Mathieson, I., I. Lazaridis, N. Rohland, S. Mallick, N. Patterson, S. A. Roodenberg, E. Harney, K. Stewardson, D. Fernandes, M. Novak, K. Sirak, C. Gamba, E. R. Jones, B. Llamas, S. Dryomov, J. Pickrell, J. L. Arsuaga, J. M. Bermúdez de Castro, E. Carbonell, F. Gerritsen, A. Khokhlov, P. Kuznetsov, M. Lozano, H. Meller, O. Mochalov, V. Moiseyev, M. A. Rojo Guerra, J. Roodenberg, J. M. Vergès, J. Krause, A. Cooper, K. W. Alt, D. Brown, D. Anthony, C. Lalueza-Fox, W. Haak, R. Pinhasi, and D. Reich. 2015. Genome-wide patterns of selection in 230 ancient Eurasians. *Nature* 528: 499–503.

Milligan, B. G. 2003. Maximum-likelihood estimation of relatedness. *Genetics* 163: 1153–1167.

Mittnik, A., K. Massy, C. Knipper, F. Wittenborn, R. Friedrich, S. Pfrengle, M. Burri, N. Carlichi-Witjes, H. Deeg, A. Furtwängler, M. Harbeck, K. von Heyking, C. Kociumaka, I. Kucukkalipci, S. Lindauer, S. Metz, A. Staskiewicz, A. Thiel, J. Wahl, W. Haak, E. Pernicka, S. Schiffels, P. W. Stockhammer, and J. Krause. 2019. Kinship-based social inequality in Bronze Age Europe. *Science* 366: 731–734.

Moltke, I., and A. Albrechtsen. 2014. RelateAdmix: A software tool for estimating relatedness between admixed individuals. *Bioinformatics* 30: 1027–1028.

Nothnagel, M., J. Schmidtke, and M. Krawczak. 2010. Potentials and limits of pairwise kinship analysis using autosomal short tandem repeat loci. *International Journal of Legal Medicine* 124: 205–215.

Olalde, I., S. Brace, M. E. Allentoft, I. Armit, K. Kristiansen, T. Booth, N. Rohland, S. Mallick, A. Szécsényi-Nagy, A. Mittnik, E. Altena, M. Lipson, I. Lazaridis, T. K. Harper, N. Patterson, N. Broomandkhoshbacht, Y. Diekmann, Z. Faltyskova, D. Fernandes, M. Ferry, E. Harney, P. de Knijff, M. Michel, J. Oppenheimer, K. Stewardson, A. Barclay, K. W. Alt, C. Liesau, P. Ríos, C. Blasco, J. V. Miguel, R. M. García, A. A. Fernández, E. Bánffy, M. Bernabò-Brea, D. Billoin, C. Bonsall, L. Bonsall, T. Allen, L. Büster, S. Carver, L. C. Navarro, O. E. Craig, G. T. Cook, B. Cunliffe, A. Denaire, K. E. Dinwiddy, N. Dodwell, M. Ernée, C. Evans, M. Kuchařík, J. F. Farré, C. Fowler, M. Gazenbeek, R. G. Pena, M. Haber-Uriarte, E. Haduch, G. Hey, N. Jowett, T. Knowles, K. Massy, S. Pfrengle, P. Lefranc, O. Lemercier, A. Lefebvre, C. H. Martínez, V. G. Olmo, A. B. Ramírez, J. L. Maurandi, T. Majó, J. I. McKinley, K. McSweeney, B. G. Mende, A. Modi, G. Kulcsár,

V. Kiss, A. Czene, R. Patay, A. Endrődi, K. Köhler, T. Hajdu, T. Szeniczey, J. Dani, Z. Bernert, M. Hoole, O. Cheronet, D. Keating, P. Velemínský, M. Dobeš, F. Candilio, F. Brown, R. F. Fernández, A. M. Herrero-Corral, S. Tusa, E. Carnieri, L. Lentini, A. Valenti, A. Zanini, C. Waddington, G. Delibes, E. Guerra-Doce, M. Neil, M. Brittain, M. Luke, R. Mortimer, J. Desideri, M. Besse, G. Brücken, M. Furmanek, A. Hałuszko, M. Mackiewicz, A. Rapiński, S. Leach, I. Soriano, K. T. Lillios, J. L. Cardoso, M. P. Pearson, P. Włodarczak, T. D. Price, P. Prieto, P. J. Rey, R. Risch, M. A. R. Guerra, A. Schmitt, J. Serralongue, A. M. Silva, V. Smrčka, L. Vergnaud, J. Zilhão, D. Caramelli, T. Higham, M. G. Thomas, D. J. Kennett, H. Fokkens, V. Heyd, A. Sheridan, K. G. Sjögren, P. W. Stockhammer, J. Krause, R. Pinhasi, W. Haak, I. Barnes, C. Lalueza-Fox, and D. Reich. 2018. The Beaker phenomenon and the genomic transformation of northwest Europe. *Nature* 555: 190–198.

Orlando, L., M. T. Gilbert, and E. Willerslev. 2015. Reconstructing ancient genomes and epigenomes. *Nature Reviews Genetics* 16: 395–408.

O'Sullivan, N., C. Posth, V. Coia, V. J. Schuenemann, T. D. Price, J. Wahl, R. Pinhasi, A. Zink, J. Krause, and F. Maixner. 2018. Ancient genome-wide analyses infer kinship structure in an Early Medieval Alemannic graveyard. *Science Advances* 4: eaao1262.

Purcell, S., B. Neale, K. Todd-Brown, L. Thomas, M. A. R. Ferreira, D. Bender, J. Maller, P. Sklar, P. I. W. de Bakker, M. J. Daly, and P. C. Sham. 2007. PLINK: A tool set for whole-genome association and population-based linkage analyses. *American Journal of Human Genetics* 81: 559–575.

Queller, D. C., and K. F. Goodnight. 1989. Estimating relatedness using genetic markers. *Evolution* 43: 258–275.

Ramstetter, M. D., T. D. Dyer, D. M. Lehman, J. E. Curran, R. Duggirala, J. Blangero, J. G. Mezey, and A. L. Williams. 2017. Benchmarking relatedness inference methods with genome-wide data from thousands of relatives. *Genetics* 207: 75–82.

Rasmussen, M., Y. Li, S. Lindgreen, J. S. Pedersen, A. Albrechtsen, I. Moltke, M. Metspalu, E. Metspalu, T. Kivisild, R. Gupta, M. Bertalan, K. Nielsen, M. T. Gilbert, Y. Wang, M. Raghavan, P. F. Campos, H. M. Kamp, A. S. Wilson, A. Gledhill, S. Tridico, M. Bunce, E. D. Lorenzen, J. Binladen, X. Guo, J. Zhao, X. Zhang, H. Zhang, Z. Li, M. Chen, L. Orlando, K. Kristiansen, M. Bak, N. Tommerup, C. Bendixen, T. L. Pierre, B. Grønnow, M. Meldgaard, C. Andreasen, S. A. Fedorova, L. P. Osipova, T. F. Higham, C. B. Ramsey, T. V. Hansen, F. C. Nielsen, M. H. Crawford, S. Brunak, T. Sicheritz-Pontén, R. Villems, R. Nielsen, A. Krogh, J. Wang, and E. Willerslev. 2010. Ancient human genome sequence of an extinct Palaeo-Eskimo. *Nature* 463: 757–762.

Reich, D. 2018. *Who we are and how we got here*. Oxford: Oxford University Press.

Renaud G., K. Hanghøj, T. S. Korneliussen, E. Willerslev, and L. Orlando. 2019. Joint estimates of heterozygosity and runs of homozygosity for modern and ancient samples. *Genetics* 212: 587–614.

Rogaev, E. I., A. P. Grigorenko, Y. K. Moliaka, G. Faskhutdinova, A. Goltsov, A. Lahti, C. Hildebrandt, E. L. W. Kittler, and I. Morozova. 2009. Genomic identification in the historical case of the Nicholas II royal family. *PNAS* 106: 5258–5263.

Russo, M. G., F. Mendisco, S. A. Avena, C. B. Dejean, and V. Seldes. 2016. Pre-Hispanic mortuary practices in Quebrada de Humahuaca (North-Western Argentina): Genetic relatedness among individuals buried in the same grave. *Annals of Human Genetics* 80: 210–220.

Saag, L., M. Laneman, L. Varul, M. Malve, H. Valk, M. A. Razzak, I. G. Shirobokov, V. I. Khartanovich, E. R. Mikhaylova, A. Kushniarevich, C. L. Scheib, A. Solnik, T. Reisberg, J. Parik, L. Saag, E. Metspalu, S. Rootsi, F. Montinaro, M. Remm, R. Mägi, E. D'Atanasio, E. R. Crema, D. Diez-del-Molino, M. G. Thomas, A. Kriiska, T. Kivisild, R. Villems, V. Lang, M. Metspalu, and K. Tambets. 2019. The arrival of Siberian ancestry connecting the Eastern Baltic to Uralic speakers further East. *Current Biology* 29: 1701–1711.

Sánchez-Quinto, F., H. Malmström, M. Fraser, L. Girdland-Flink, E. M. Svensson, L. G. Simões, R. George, N. Hollfelder, G. Burenhult, G. Noble, K. Britton, S. Talamo, N. Curtis, H. Brzobohata, R. Sumberova, A. Götherström, J. Storå, and M. Jakobsson. 2019. Megalithic tombs in western and northern Neolithic Europe were linked to a kindred society. *PNAS* 116: 9469–9474.

Santana, J., F. J. Rodríguez-Santos, M. D. Camalich-Massieu, D. Martín-Socas, and R. Fregel. 2019. Aggressive or funerary cannibalism? Skull-cup and human bone manipulation in Cueva de El Toro (Early Neolithic, southern Iberia). *American Journal of Physical Anthropology* 169: 31–54.

Scheib, C. L., R. Hui, E. D'Atanasio, A. W. Wohns, S. A. Inskip, A. Rose, C. Cessford, T. C. O'Connell, J. E. Robb, C. Evans, R. Patten, and T. Kivisild. 2019. East Anglian early Neolithic monument burial linked to contemporary Megaliths. *Annals of Human Biology* 46: 145-149.

Schroeder, H., A. Margaryan, M. Szmyt, B. Theulot, P. Włodarczak, S. Rasmussen, S. Gopalakrishnan, A. Szczepanek, T. Konopka, T. Z. T. Jensen, B. Witkowska, S. Wilk, M. M. Przybyła, Ł. Pospieszny, K.-G. Sjögren, Z. Belka, J. Olsen, K. Kristiansen, E. Willerslev, K. M. Frei, M. Sikora, N. N. Johannsen, and M. E. Allentoft. 2019. Unraveling ancestry, kinship, and violence in a Late Neolithic mass grave. *PNAS* 116: 10705–10710.

Sikora, M., A. Seguin-Orlando, V. C. Sousa, A. Albrechtsen, T. Korneliussen, A. Ko, S. Rasmussen, I. Dupanloup, P. R. Nigst, M. D. Bosch, G. Renaud, M. E. Allentoft, A. Margaryan, S. V. Vasilyev, E. V. Veselovskaya, S. B. Borutskaya, T. Deviese, D. Comeskey, T. Higham, A. Manica, R. Foley, D. J. Meltzer, R. Nielsen, L. Excoffier, M. M. Lahr, L. Orlando, and E. Willerslev. 2017. Ancient genomes show social and reproductive behavior of early Upper Paleolithic foragers. *Science* 358: 659–662. doi: 10.1126/science.aao1807

Snyder-Mackler, N., W. H. Majoros, M. L. Yuan, A. O. Shaver, J. B. Gordon, G. H. Kopp, S. A. Schlebusch, J. D. Wall, S. C. Alberts, S. Mukherjee, X. Zhou, and J. Tung. 2016. Efficient genome-wide sequencing and low-coverage pedigree analysis from noninvasively collected samples. *Genetics* 203: 699–714.

Speed, D., and D. J. Balding. 2015. Relatedness in the post-genomic era: Is it still useful? *Nature Reviews Genetics* 16: 33–44.

Stevens, E. L., G. Heckenberg, E. D. O. Roberson, J. D. Baugher, T. J. Downey, and J. Pevsner. 2011. Inference of relationships in population data using identity-by-descent and identity-by-state. *PLoS Genetics* 7: e1002287.

The 1000 Genomes Project Consortium. 2015. A global reference for human genetic variation. *Nature* 526: 68–74.

Theunert, C., F. Racimo, and M. Slatkin. 2017. Joint estimation of relatedness coefficients and allele frequencies from ancient samples. *Genetics* 206: 1025–1035.

Thornton, T., H. Tang, T. J. Hoffmann, H. M. Ochs-Balcom, B. J. Caan, and N. Risch. 2012. Estimating kinship in admixed populations. *American Journal of Human Genetics* 91: 122–138.

Trinkaus, E., A. P. Buzhilova, M. B. Mednikova, and M. V. Dobrovolskaya. 2014. *The people of Sunghir*. New York: Oxford University Press.

Villalba-Mouco, V., M. S. van de Loosdrecht, C. Posth, R. Mora, J. Martínez-Moreno, M. Rojo-Guerra, D. C. Salazar-García, J. I. Royo-Guillén, M. Kunst, H. Rougier, I. Crevecoeur, H. Arcusa-Magallón, C. Tejedor-Rodríguez, I. García-Martínez de Lagrán, R. Garrido-Pena, K. W. Alt, C. Jeong, S. Schiffels, P. Utrilla, J. Krause, and W. Haak. 2019. Survival of Late Pleistocene hunter-gatherer ancestry in the Iberian Peninsula. *Current Biology* 29: 1169–1177.

Wang, J. 2011a. COANCESTRY: A program for simulating, estimating and analysing relatedness and inbreeding coefficients. *Molecular Ecology Resources* 11: 141–5.

Wang, J. 2011b. Unbiased relatedness estimation in structured populations. *Genetics* 187: 887–901.

Wang, J. 2014. Marker-based estimates of relatedness and inbreeding coefficients: An assessment of current methods. *Journal of Evolutionary Biology* 27: 518–530.

Wang, C.-C., S. Reinhold, A. Kalmykov, A. Wissgott, G. Brandt, C. Jeong, O. Cheronet, M. Ferry, E. Harney, D. Keating, S. Mallick, N. Rohland, K. Stewardson, A. R. Kantorovich, V. E. Maslov, V. G. Petrenko, V. R. Erlikh, B. C. Atabiev, R. G. Magomedov, P. L. Kohl, K. W. Alt, S. L. Pichler, C. Gerling, H. Meller, B. Vardanyan, L. Yeganyan, A. D. Rezepkin, D. Mariaschk, N. Berezina, J. Gresky, K. Fuchs, C. Knipper, S. Schiffels, E. Balanovska, O. Balanovsky, I. Mathieson, T. Higham, Y. B. Berezin, A. Buzhilova, V. Trifonov, R. Pinhasi, A. B. Belinskij, D. Reich, S. Hansen, J. Krause, and W. Haak. 2019. Ancient human genome-wide data from a 3000-year interval in the Caucasus corresponds with eco-geographic regions. *Nature Communications* 10: 590.

Waples, R., A. Albrechtsen, and I. Moltke. 2019. Allele frequency-free inference of close familial relationships from genotypes or low-depth sequencing data. *Molecular Ecology* 28: 35–48.

Weir, B. S., A. D. Anderson, and A. B. Hepler. 2006. Genetic relatedness analysis: Modern data and new challenges. *Nature Reviews Genetics* 7: 771–780.

Zhu, X., S. Li, R. S. Cooper, and R. C. Elston. 2008. A unified association analysis approach for family and unrelated samples correcting for stratification. *American Journal of Human Genetics* 82: 352–365.

CHAPTER 7

The structure of cranial morphological variance in Asia: Implications for the study of modern human dispersion across the planet

Mark Hubbe[1,2]

Abstract

The study of past human dispersion is a central topic to understand how humans occupied the planet. However, this is not a simple task, as it depends on our ability to estimate the ancestral state of past populations based on the biological diversity observed among samples, before this can be used to infer phylogenetic relationships between them. Studies dedicated to this type of analysis rely on understanding the forces that structured the variance between and within the groups studied. For complex phenotypic data, like cranial morphological variation, this task is especially challenging due to the combination of factors that contributed to the observed pattern of variance among modern human groups. In this chapter, I explore the structure of the morphological variance within and between Asian regions, to illustrate the importance of considering the myriad of evolutionary forces structuring the morphological variance among regions when reconstructing past human dispersion. Using a large craniometric dataset representing three Asian macro-regions, I compare, through the estimation of Q_{ST} values, the apportionment of variance in Asia and compare it to the values obtained for other regions of the planet. The results are contextualized within four continent-wide periods of human dispersion. Taken together, they suggest that Asia's morphological diversity is mostly defined by a geographic structure, same as the rest of the planet, but there are significant departures from this pattern when analyzing Northeast Asia. This analysis shows that local conditions can impact significantly the structure of morphological variance and must be considered in the reconstruction of past dispersion events.

INTRODUCTION

One of the main challenges faced by scientists interested in the study of modern humans' past is the reconstruction of the pathways by which

---

[1] Department of Anthropology, Ohio State University, USA.
[2] Instituto de Arqueología y Antropología, Universidad Católica del Norte, Chile.

© 2021, Kerns Verlag / https://doi.org/https://doi.org/10.51315/9783935751377.007
Cite this article: Hubbe, M. 2021. The structure of cranial morphological variance in Asia: Implications for the study of modern human dispersion across the planet. In *Ancient Connections in Eurasia*, ed. by H. Reyes-Centeno and K. Harvati, pp. 129-156. Tübingen: Kerns Verlag. ISBN: 978-3-935751-37-7.

human populations dispersed across different geographic areas over time. Hundreds of articles have been published in the last decade alone dedicated to the discussion about the colonization of and dispersion over larger continental areas, testing and/or defending different migration hypotheses or dispersion scenarios. There is no doubt that this type of inquiry has been at the core of our efforts to understand the origin of modern human biological and cultural diversity. The most fundamental challenge for these studies lies in the fact that, invariably, we are reconstructing past populations dynamics by studying samples that do not directly belong to the populations that were involved in the migration or dispersion process; they only represent the descendants of the individuals involved in those dispersion events. Even studies working with prehistoric remains rarely—if ever—are able to assess the characteristics of the populations that were involved in the demic diffusion process of interest. As such, the study of past human dispersion events relies on the analysis of the structure of variance observed in descendant populations, more specifically on the partitioning of the observed variance that results from differences between-groups and of individual variations within-groups.

The structure of variance between and within-groups has been the basis of studies about past human dispersion events because its analysis permits the inference of the origins of such variance and, consequently, the reconstruction of the probable biological history of the populations sampled. For example, differences between groups (also referred to biological distances or biodistances) have been frequently used to explore biological affinities between populations in the past (e.g., Hanihara 1996; Howells 1989; Neves and Hubbe 2005) and have also been used to test specific hypotheses of possible dispersion models. Frequently, this is done by associating such biological distances with specific evolutionary forces causing them, like for example patterns of gene flow over time or space (e.g., Harvati and Weaver 2006; Hubbe, Harvati, and Neves 2011; Reyes-Centeno et al. 2015; von Cramon-Taubadel 2009; von Cramon-Taubadel, Strauss, and Hubbe 2017). Ultimately, however, the common denominator in all these studies is the assumption (implicit or explicit) that the evolutionary forces structuring the variance observed in the populations sampled are well understood, and therefore they allow for the derivation of the population's ancestral characteristics (i.e., their evolutionary history).

Therefore, the structure of human phenotypic (and genetic) variance between and within-populations is a key aspect in the study of past human population dynamics (e.g., Relethford 1994; Roseman 2004; von Cramon-Taubadel 2009; von Cramon-Taubadel and Weaver 2009), as it forms the basis for any exploratory or model bound analysis of past population movements, and informs the ways by which human groups occupied and adapted to different geographic regions in the planet. Succinctly, the structure of phenotypic variance results from different evolutionary forces that regulate the origin, spread, and maintenance of variance over time. In the case of complex phenotypic traits, such as cranial morphology, the way by which specific genetic, developmental and environmental factors interact to produce the final cranial shape of any given individual is still largely unknown (e.g., Hallgrímsson et al. 2007; Klingenberg 2014), but the structure of the variance among and within groups (i.e., the relative distances among them) has been shown to be correlated to (and in some cases to be the result of) several quantifiable phenomena, ranging from stochastic evolutionary processes, like genetic drift and isolation by distance, to natural selection to specific environmental pressures or even cultural practices that regulate gene flow between populations (e.g., Galland et al. 2016; González-José et al. 2005; von Cramon-Taubadel, 2014). These phenomena, when combined, result in specific patterns of partitioning of the variance observed between and within-populations, resulting in the relative biological differences between them. The differences (i.e., biological distances) between groups are then used as theoretical assumptions in studies focused on understanding the biological affinities and relationships among populations in the past.

The assumptions derived from this approach are frequently used to test null hypotheses, in which the biological distance represents the expected divergence from a common ancestor under random evolutionary processes like genetic drift. In that way, the departure from expectation of these neutral null hypotheses permits the discussion of other factors that may explain the distances observed. For instance, there is a vast literature dealing with the reliability of using cranial morphological differences to reconstruct major aspects of human dispersion across the planet (e.g., Betti et al. 2009; Carson 2006; Harvati and Weaver 2006; Hubbe, Hanihara, and Harvati 2009; Relethford 1994, 2004; Roseman 2004; Smith 2009; von Cramon-Taubadel and Weaver 2009), given the importance of these data to assess the biological characteristics of populations from regions or timeframes of interest, especially in cases where access to direct genetic information is limited. This discussion has been particularly present in the last couple of decades, as the study of morphological affinities has become a central component in the study of past human mobility, playing a major role in the discussions about the human occupation of Asia (e.g., Hanihara 1996; Harvati 2009; Reyes-Centeno et al. 2015), Europe (e.g., Pinhasi and von Cramon-Taubadel 2012), Australo-Melanesia (e.g., Schillaci 2008), the Americas (e.g., de Azevedo et al. 2011; Strauss et al. 2015; von Cramon-Taubadel et al. 2017), and Polynesia (e.g., Valentin et al. 2016), not to mention studies focused on smaller geographical regions. Understanding the evolutionary forces that structure the accumulation and maintenance of phenotypic differences between populations is therefore essential to build well-informed models and hypotheses that can be tested with morphological data. Understandably, a stronger knowledge about the evolutionary forces responsible for the partitioning of variance between populations leads to stronger predictive models, which in turn can be used to construct more accurate explanations for interpreting the patterns of human dispersion in the past. Therefore, resolving the source of the variance partitioning, including the interactions between genotype, environment, development and phenotype, among modern humans is a necessary step for the study of past human dispersion.

Evolutionary forces shaping morphological variance

The discussion about the sources of cranial morphological variance partitioning has been especially focused on the role that non-directional evolutionary forces have played in shaping the modern human morphological variation worldwide. While the structure of variance between and within populations depends both on the development of new variance (for example, through mutations or admixture with other demes) and on the redistribution of the variance (through, for example, gene flow and genetic drift), the discussion about the structure of cranial morphological variance among modern humans has been focused almost exclusively on the latter. Given the developmental constrains during the growth of cranial structures, as well as the different degrees of integration between its anatomical modules (see review in Lieberman 2011), the acquisition of new variance through mutations is considered negligible when compared to the magnitude of differences that can be accumulated from the unequal redistribution of the available variance across populations. For these reasons, most of the studies focused on the origins of morphological variance among modern humans have concentrated on distinguishing neutral from non-neutral micro-evolutionary forces (e.g., Betti et al. 2009; Harvati and Weaver 2006; Hubbe et al. 2009; Relethford 2004; Roseman 2004; Smith 2009).

Some of the most impactful studies in this sense are those that support the idea that morphological variance is structured according to stochastic events, like serial bottlenecks or isolation by distance. Most of these studies compare morphological data to the patterns of variance observed in neutral molecular markers or with specific predictions of differentiation over time and space that derive from neutral evolutionary models (e.g., Betti et al. 2009; Relethford 2004; Roseman 2004; von Cramon-Taubadel and Weaver 2009). What makes these studies so relevant for the reconstruction of past human dispersion is the fact that they allow to quantify the impact of evolutionary forces structuring morphological variance (and distances between-groups) that are relatively linear over time and space. In other words, they allow reconstructing the ancestral conditions of the populations studied through linear relationships that are not dependent on specific extrinsic factors. As long as the morphological distances between populations are largely the result of stochastic events, it is expected that geographic or temporal distances between series will be linearly correlated with gene flow between populations, since gene flow will be reduced over space and time through a combination of events like isolation by distance or serial founder effects. As this depends only on intrinsic factors of the sample studied (i.e., the way by which total variance is structured between and within populations), several studies have used this assumption to test dispersion scenarios among past populations (e.g., Hubbe, Neves, and Harvati 2010; Pinhasi and von Cramon-Taubadel 2012; Reyes-Centeno et al. 2015; Strauss et al. 2015).

Non-stochastic factors, on the other hand, pose a significant challenge to studies of past human dispersions. Non-stochastic factors, like natural selection to extremely cold environments (Harvati and Weaver 2006; Hubbe et al. 2009), or adaptive responses to changes in diet and subsistence (Galland et al. 2016; González-José et al. 2005) have been identified in several regions of the planet and are responsible for significant portions of the cranial phenotypic variance partitioning (i.e., relative distances among populations). However, while they are undeniably important components of the origin of modern human morphological variance, their context-dependent nature makes them harder, if not impossible, to be included in models of past human dispersion processes. Consequently, the magnitude and direction of responses under non-stochastic forces are dependent largely on factors that are not intrinsic to the populations studied, like environmental oscillations, technological innovations or even social constrains, and these factors are rarely known in enough detail to permit their incorporation in models that infer the ancestral condition of the populations studied. These factors limit significantly our ability to create accurate models of dispersion across space and time, since they limit the reconstruction of the ancestral states of populations, or the time and distance that separate modern populations from last common ancestors. In other words, they break the assumption of linear accumulation of differences between populations since their last common ancestor and can result in incorrect reconstructions of population history. Therefore, when biological distances are used to reconstruct routes of population dispersion, the presence of non-stochastic events affecting some of the clades in the analysis can result in the evolutionary relationship of populations being incorrectly quantified. For example, under strong stabilizing selection, clades will remain more similar over time than expected under neutral models, while under directional selection, clades will accumulate differences faster over time than under stochastic events. Unless properly quantified, this departure from linear associations between biological differences in ancestral-descendant relationships can generate distorted reconstruction of past events, by suggesting relationships between groups that are artificial and that may mask the real evolutionary history between populations. Although non-stochastic factors have been argued to be of less importance on larger geographic scales (e.g., Evteev et al. 2014; Relethford 2004; Roseman 2004; von Cramon-Taubadel 2011), several studies in the past have claimed that they can be important enough as to deem studies of biological affinities between populations unreliable (e.g., Carlson and Van Gerven 1979).

Similar to non-stochastic evolutionary forces, other factors can disrupt the linear expectation between biological distances and ancestor-descendant relationships. Of particular interest to the context of this chapter are disruptions caused by multiple dispersion events into specific regions, where the latter migrations may partially or completely erase the previous dispersions. In these cases, early dispersion events can be completely invisible from the morphological data (in cases of full replacement), or can add considerable noise to the biological distances among populations (in case of partial assimilation of earlier populations' variance), and therefore also produce artificial evolutionary relationships between the populations analyzed. Despite the fact that multiple dispersion events characterize the occupation of most, if not all, larger regions of the planet, most studies of morphological diversity among modern humans tend to assume that the process of human dispersion is the result of a single dispersion wave or that the geneflow between populations was constant over time and space. This is a necessary assumption in most cases, since the incorporation of multiple dispersion events requires knowledge that is extrinsic to the populations studied, and that varies from region to region, similar to what happens in cases where non-stochastic evolutionary forces acted on morphological diversity.

The main point of this brief discussion is to illustrate that understanding how the variance among modern humans is structured (i.e., what is producing the biological distances observed among them) is a central aspect to reconstruct past population dispersion events, which at the same time is essential for reconstructing the pathways by which our species expanded and occupied the planet. Even if most studies do not acknowledge this explicitly, all studies of human dispersion over space and time depend on accurate reconstructions of ancestral states of the populations being studied to establish the last common ancestor between groups of interest, which in turn allows establishing phylogenetic relationships between them. Only when accurate phylogenetic relationships have been established, can this information be contrasted with the spatial or temporal characteristics of the studied series to infer the dispersion patterns that gave origin to them. In other words, they all rely on accurate assumptions of evolutionary forces that structured the observed variance between and within populations.

In this chapter, I will explore the structure of the cranial morphological variance in Asia, contrasting it with what is observed in other continents and macro-regions, to contribute to the discussion about the evolutionary forces structuring morphological variance among modern human populations. I have two main goals with this analysis. The first one is to test the hypothesis that modern human morphological variation is structured similarly across the planet, i.e., that we can assume that the relationship between morphological distances and ancestral-descendant relationships are largely linear across the planet. As introduced above, this is a central assumption in several studies that explore the morphological diversity across larger regions of the planet, and a better understanding of the similarities and differences in the ways that variance is partitioned across regions can contribute to studies of past human migration.

My second goal is to look more specifically at the impact that different evolutionary forces and dispersion events played in the partitioning of morphological variance among different regions of Asia, in order to contribute to our understanding of how much the unique history of Asia's human occupation is reflected in the larger pattern of human morphological diversity. For this end, I derive some expectations from four major events of dispersion that characterize the human presence in Asia, and contrast them to the partitioning of variance among Asian regions. I focus on four major stages that may have caused widely spread and long-lasting impacts on the structure of Asian morphological variance (i.e., on the pattern of relative biological distances among populations). These stages are evidently simplistic and clearly cannot be considered as an exhaustive review of the human dispersion across Asia. They were selected because they can be translated into predictions to be contrasted against the craniometric data available for this study.

Modern human dispersions across Asia

Asia is an interesting continent to focus for the analysis of modern human dispersion for a series of complementary reasons. First, it represents the largest continent on the planet, encompassing around 30% of the world's landmasses, and presents a wide range of ecological zones, several of them (e.g., subarctic and high-altitude areas) far outside the original conditions under which our species evolved. Second, the history of the human occupation and dispersion across the Asian continent has been a major focus of recent archaeological, anthropological, linguistic and genetic research, as well illustrated by the contributions presented in this volume. Moreover, the human dispersion across Asia is of relevance not only for understanding the processes that allowed human population to successfully colonize the immense ecological diversity of the continent, but also because the occupation of Asia played a key role in the human dispersion across the planet during the end of the Pleistocene. And finally, Asia went through several stages of broad human dispersion that may have changed the structure of the morphological diversity among modern human populations over time across large portions of the continent. As such, the process of human occupation of Asia's vastly different geographic and ecological regions was marked by complex patterns of mobility and interaction between human groups in the past.

Broadly speaking, Asia's history can be divided into four large moments of population expansion, which have contributed to the modern biological makeup of Asian populations. Each of them represents events that are unique to the continent, and a better understanding of them allows to answers the question of how much of the morphological variance seen in Asia is structured in ways similar to what is observed in other regions of the planet, and how much of it is unique to this continent.

The first main stage of human dispersion in Asia is represented by the original expansion of modern humans from Africa into this continent.

Asia was occupied by modern humans very early (~130-70 kyr; Lahr 1996; Mellars 2006; Mellars et al. 2013; Reyes-Centeno et al. 2014) when compared to the occupation of other non-African continents. The initial dispersion into Asia kept the human populations within the environmental range of tropical and subtropical regions, possibly following a coastal route (Petraglia et al. 2010; Reyes-Centeno et al. 2014), which promoted an eastward dispersion. This initial wave out of Africa is marked by a strong initial genetic bottleneck (Betti et al. 2009; Manica et al. 2007), followed by serial founder-effects, resulting in a loss of average genetic and morphological diversity as populations colonized regions farther away from Africa. However, early Asian populations met and probably admixed with multiple hominin groups that were previously inhabiting Asia (Teixeira and Cooper 2019), resulting in a potential influx of genetic variance back into Asian populations. From a cranial morphological point of view, the early populations expanding out of Africa seem to have retained the same ancestral morphological patterns of the earliest African modern humans, so early Asian specimens tend to be closer morphologically to Late Pleistocene specimens in Africa, Europe and Australia (Harvati 2009; Hubbe et al. 2011) than to most modern Asian populations. This earlier morphological pattern was mostly replaced during later periods, although isolated populations throughout Southeast Asia have been described as retaining the earlier morphology (Lahr 1996; Reyes-Centeno et al. 2015). Although this stage represents an important step in the human occupation of Asia, the evidence that only a few of the modern populations are their direct descendants suggests that there will not be a strong impact of this stage in the structure of variance observed among recent Asian populations. This initial dispersion event follows the expected pattern of isolation by distance and serial bottle-necks, and as such would be expected to leave similar signatures in the apportionment of variance as would be seen in other larger areas occupied by humans in the past. However, it is possible that admixture with hominin groups already occupying Asia during this period would result in an influx of genetic diversity into modern human populations residing where these encounters happen. Moreover, if they were frequent enough, this would result in a deviation from linear patterns, as a result of morphological differences between groups increasing faster than what would be expected under stochastic models.

The second stage of occupation is associated with subsequent dispersion waves out of Africa, which followed the initial colonization of the continent. While the suggestion of multiple dispersions into Asia is not supported by some autosomal studies (HUGO Pan-Asian SNP Consortium et al. 2009; Reich et al. 2011), several archaeological, genetic, and craniometric studies (Mellars 2006; Rasmussen et al. 2011; Reyes-Centeno et al. 2014) find support for this scenario. This second stage of dispersion may have followed quickly after the first one, starting as early as 50 thousand years ago, and is associated with both an eastward path across the continent and probably with the occupation of more temperate zones. The expansion waves following the initial occupation of Asia seem to have remained largely separated from earlier populations, as there is little evidence of admixture between them (Rasmussen et al. 2011; Reyes-Centeno et al. 2014). While there is not enough archaeological evidence at this point to refine the chronological timeframe of this expansion, or even to answer questions about the routes taken by them (see Reyes-Centeno et al. 2014; Reyes-Centeno et al. 2015 for a longer discussion), this expansion wave occupied larger portions of the continent, following the same pattern of serial founder-effects and isolation by distance from Africa proposed for the earliest expansions into the continent. Probably, this is the event that shaped most strongly the morphological diversity seen in the Asian continent, given the geographic structure of modern Asian populations (HUGO Pan-Asian SNP Consortium et al. 2009), and as such is the stage that is expected to have the largest impact on the current structure of Asian cranial morphological variance.

A third important stage in the human dispersion across Asia relates to the occupation of higher latitudes, which may have started as early as 35 kyr (e.g., Jacobs et al. 2019). The expansion into higher latitudes and colder environments marks an important feat for modern human populations, as it represents the expansion into ecological zones that humans are not able to survive without technological advances that permit them to conquer these harsh environments. The expansion into colder climates occurred much later than the initial occupation of Australia, for example, and was also the last step required before humans could settle the Americas and occupy all the large continents in the planet. Such challenges are reflected as well in the genetic diversity of northern Asian populations, since there is a clear clinal structure of decreasing genetic diversity from south to north in the continent (HUGO Pan-Asian SNP Consortium et al. 2009). From the point of view of morphological variance, the dispersion towards higher latitudes is an important stage in the structuring of morphological variance, since craniometric studies have demonstrated a strong adaptive response in populations occupying the extreme cold of high latitude environments (Evteev et al. 2014; Harvati and Weaver 2006; Hubbe et al. 2009). As such, the occupation of higher latitudes is associated with the first strong departure from a purely isolation by distance pattern that is assumed to have structured the morphological variance in Asia until then. The adaptive response to cold climate promoted an increase in the morphological distances between populations, particularly in anatomical regions of the skull associated with the regulation of internal body temperature, like the neurocranium and the nasal region (Hubbe et al. 2009).

The final stage that is important in the context of this chapter is the increased mobility and gene flow experienced in large portions of Asia during the second half of the Holocene. The domestication of horses, the creation of long-distance trade routes, like the Silk Road, and the spread of the larger civilizations in East and South Asia may have had a significant impact in increasing gene flow across Asia, decreasing the amount of variance that is a result of differences between groups (i.e., making populations more similar to each other). This larger degree of connection across Asia is supported by historical and linguistic documentation (e.g., Hansen 2012), as well as genetic evidence. Although there is a clear pattern of genetic structure based on geography and language groups in the continent (HUGO Pan-Asian SNP Consortium et al. 2009), a large proportion of Asian Y-chromosome diversity has been shown to derive from a very small number of male individuals who lived between 4 and 2 kyr BP (Balaresque et al. 2015; Zerjal et al. 2003), supporting the quick spread of genetic lineages across Asia. The expectation that is derived from this period of increased population movement is that some of the previous variance structure will be diluted or even erased, as gene flow across large distances will diminish the amount of variance that is a result of differences between groups.

Together, these four stages combine several different processes associated with human dispersion across the continent and illustrate how population movement may have affected the current structure of local morphological variance. While some of them probably acted in conjunction, structuring variance between and within groups in similar ways (stages 1 and 2), others acted counter to this initial structure, creating possible deviations associated with increased variance (stage 3) or decreasing distances between groups (stage 4). As such, these stages create a series of expectations that can be tested against the morphological data analyzed in this chapter:

1. The first stage of human dispersion would have created a strong geographic structure to morphological variance in the continent, due to serial founder effects and bottlenecks from Africa. This would result in strong correlations between morphological distances and geographic distances. On the other hand, this is the stage where admixture with other hominins population was probably most frequent, which could inflate the amount of variance that is due to differences between groups. However, it is also possible that the signal form this dispersion event was largely erased by the subsequent dispersions of the second stage.
2. The second stage of human dispersion would also be highly geographically structured and would result in the morphological diversity of Asia reflecting patterns seen in other parts of the planet, following the expectations from isolation by distance and sequential bottleneck processes.
3. The third stage probably resulted in a strong climate signal on the structure of variance in Asia, as natural selection to cold environments would result in inflated morphological distances between populations from cold and warmer environments.
4. Finally, the fourth stage acted as a homogenizing event, decreasing the differences among populations as gene-flow was facilitated by

technological advancements and large demic expansions late in the Holocene.

MATERIAL AND METHODS: QUANTIFYING CRANIAL MORPHOLOGICAL VARIANCE IN ASIA

To explore the morphological variance of recent Asian populations, I used a dataset of 33 linear craniometric measurements from 7422 individuals collected by Prof. Tsuheniko Hanihara, representing 135 populations from all major regions of the planet (Table 1). Details of the dataset can be obtained in Hanihara (1996) and Hubbe et al. (2009). Table 2 presents a breakdown of the total sample by geographic regions and show how the different series were grouped together. The series represent samples of recent pre-industrial populations and are composed only of male individuals, because the sample size for females is considerably smaller and would preclude the analysis of a similar number of populations. The series were grouped by continent and also according to 14 subcontinental macro-regions plus a region representing Polynesia, that were defined based on general geographic and ecological proximity (Fig. 1; Hubbe et al. 2009). Asia was subdivided into three macro-regions —South, East and Northeast Asia—based on the series available.

The definition of the macro-regions followed the geographic distribution of the available samples as well as the general geographic proximities between them. While these groupings are not based on the biological history of the populations included in them, and therefore may represent artificial separations between them, previous studies (Hanihara 1996; Hubbe et al. 2009) have shown strong consistency among the series

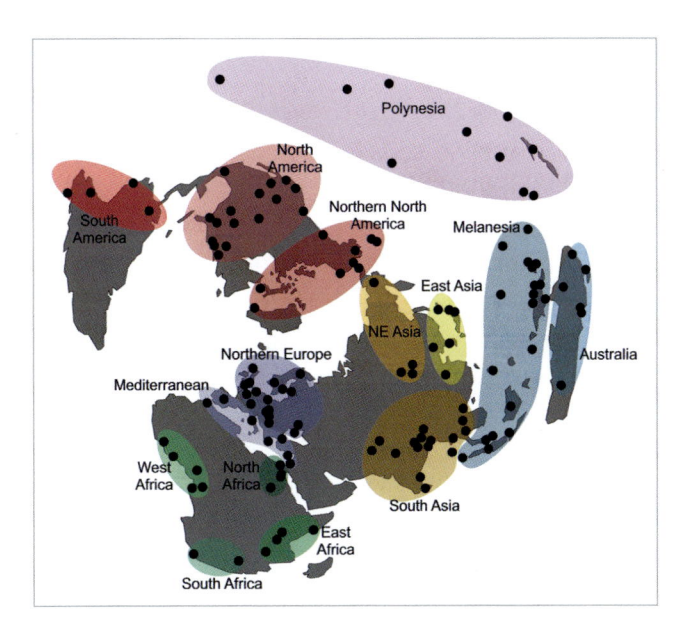

Fig. 1.
Geographic location of series included in the study and visual representation of the 15 geographic macro-regions analyzed.

Table 1.
Linear measurements included in this study.

Variables	Reference
Maximum cranial length (GOL)	Howells 1973
Nasion-opisthocranion (NOL)	Howells 1973
Cranial base length (BNL)	Howells 1973
Maximum cranial breadth (XCB)	Howells 1973
Minimum frontal breadth (M9)	Martin and Saller 1957
Maximum frontal breadth (XFB)	Howells 1973
Biauricular breadth (AUB)	Howells 1973
Biasterionic breadth (ASB)	Howells 1973
Basion –bregma height (BBH)	Howells 1973
Sagittal frontal arc (M26)	Martin and Saller 1957
Saggital parietal arc (M27)	Martin and Saller 1957
Saggital occipital arc (M28)	Martin and Saller 1957
Nasion-bregma chord (FRC)	Howells 1973
Bregma-lambda chord (PAC)	Howells 1973
Lambda-opisthion chord (OCC)	Howells 1973
Basion prosthion length (BPL)	Howells 1973
Breadth between Frontomalare temporale (M43)	Martin and Saller 1957
Bizygomatic breadth (ZYB)	Howells 1973
Nasion prosthion height (NPH)	Howells 1973
Interorbital breadth (DKB)	Howells 1973
Orbital breadth (M51)	Martin and Saller 1957
Orbital height (OBH)	Howells 1973
Nasal breadth (NLB)	Howells 1973
Nasal height (NLH)	Howells 1973
Palate breadth (MAB)	Howells 1973
Mastoid height (MDH)	Howells 1973
Mastoid width (MDB)	Howells 1973
Frontal chord (M43(1))	Martin and Saller 1957
Frontal subtense (No 43c)	Bräuer 1988
Simotic chord (WNB)	Howells 1973
Simotic subtense (SIS)	Howells 1973
Zygomaxillary chord (ZMB)	Howells 1973
Zygomaxillary subtense (SSS)	Howells 1973

Table 2.
Sample size and composition for continents and macro-regions in the study.
* Series' names follow Hanihara (1996).

Continent	Macro-regions	Series included in regions*	Total individuals
Asia	East Asia	Han North; Manchuria; Han South; Jomon Japan; Ainu; Tohoku Japan; Tokyo Japan; Korea	404
	North East Asia	Buriat; Chuckchis; Mongols	188
	South Asia	Nepal; Tibete; Bengala; Bihar; Madras; NW India; Punjab; Sikkim; Andaman islands; Laos; Malasia; Myanmar; Singapure; Thailand; Vietnam; Afganisthan; Bangladesh	774
Australo-Melanesia	Melanesia	Fiji; New Britain; New Caledonia; New Hebrides; New Ireland; Solomon; Torres Strait; Carolina Islands; Mariana Islands; Easter Papua; Gulf Province; Madang; East Sepik; West Sepik; Borneo; Java; Molucca; Negritos Phillipines; Phillipines; Sumatra	852
	Australia	New South Wales; Queensland; South Australia; West Australia	193
Africa	East Africa	Kenya; Malawi; Somalia; Tanzania	244
	North Africa	Nubia; Early Nubia; Kerma	203
	South Africa	South Africa; Bushman; Kaffir; Zulu	125
	West Africa	Cameroon; Gabon; Ghana; Ivory Coast; Ibo Nigeria; Others Nigeria	296
Americas	North America	México; Arizona; Arkansas; North California; South California; South Dakota; Florida; Illinois; Kentucky; Maryland; New Mexico; New York; Utah; Virginia	972
	Northern North America	West Aleuts; East Aleuts; British Columbia; Iroques Ontario; SW Alaska; NW Alaska; N Alaska; NE Asia; NE Canada; Greenland; Alaska; Tlingit Alaska	720
	South America	Chile; Fueguinos; Peru; Venezuela + Colombia	317
Europe	Mediterranean	Recent Greece; Recent Italy; Ancient Italy; Spain + Portugal; Egypt; Badari Egypt; Gyzeh; Naqada; Marrocos; Iron Age Israel; Iron and Bronze Age Israel; Turkey	725
	North Europe	Austria; Czech; Finland; Recent France; Germany; Netherlands; Hungary; Lapps; Russia; Sweden; Switzerland; Ukraine; Serbia; Ensay; Poundbury; Repton; Spittafields CAM; Spittafields NHM	979
	Polynesia	Chatam Islands; Easter Island; Hawaii; Marquese Islands; New Zealand; Society Islands	430
Total		N = 135	7422

within the macro-regions, supporting their use in this study. Evidently, it is possible that more refined divisions could be defined with these data, but they would result in either a small sample of series per macro-region, or in macro-regions that overlapped. As such, the 15 macro-regions used here represent the best compromise found by the author to show the natural geographic sub-divisions of the series while maintaining enough series in each macro-region to estimate the partitioning of their variances.

This study relies on the use of linear measurements, to take full advantage of the large size of the data collected by Dr. Hanihara, which represents the largest available dataset for craniometric diversity worldwide, and includes several collections that have been repatriated recently and are not available for study anymore. Although linear measurements have been largely replaced by the analysis of 3D morphometric data in the last decades, as the latter is inherently more efficient at measuring and representing shape, Hanihara's dataset is uniquely robust to quantify the portioning of variance within and between groups on a global scale, and has been demonstrated to correlate well with other measurements of modern human diversity (Hanihara 1996). Therefore, despite the limitations of linear measurements, this particular dataset represents the best dataset available to study the structure of variance within and between modern human groups on a global scale.

Prior to the analysis of the data, the effect of size of the individuals was removed from the craniometric measurements by dividing each measurement by the geometric mean of the individuals (Darroch and Mosimann 1985; Hubbe et al. 2011). Given the focus on the structure of variance between and within-groups, I assessed the morphological variance in the dataset by using Q_{ST} estimates (or minimum Fst; Relethford 1994; Relethford and Blangero 1990), which are the approximations of genetic Fst for quantitative traits. The Q_{ST} estimates measure the proportion of the variance in a group of samples that is due to the differences between groups. They were calculated following Relethford and Blangero (1990), assuming an average heritability of 1.0 to present conservative estimates. Although craniometric traits have been shown to have only low to moderate heritability values (Carson 2006), the conservative heritability adopted here does not affect the arithmetic relationship between values estimated for different datasets, and will not change the hierarchy of values obtained, as long as the heritabilities can be assumed to be similar across all datasets being compared. Since all series here are representatives of recent modern humans, this is a reasonable assumption.

The Q_{ST} estimates were calculated for each of the continents separately and then for each of the 15 macro-regions, to explore the proportion of variance in each combination of series that is the product of differences between groups. To extend this analysis, I also applied the Relethford-Blangero analysis (Relethford and Blangero 1990) for the continental and sub-continental groups, which compares the observed variance within each series with the expected variance based on the R-matrix calculated from these data. The R matrix consists of the standardized variances and covariances of populations around the genetic centroid of the data, and it permits the calculation of the expected phenotypic variances for each group analyzed. As a result, the Relethford-Blangero analysis calculates the residuals between the observed phenotypic variance and the variance that would be expected based on the distance of each group to the genetic centroid in the data (Rii). Assuming all groups included in the analysis are part of the same macro-population, the analysis permits to infer which groups have higher (positive residuals) or lower (negative residuals) observed variances than would be expected by its distance from the genetic centroid in the data.

The structure of the variance in Asian populations was further explored by analyzing the relationship between Q_{ST} estimates and geographic distances between series. Under a null hypothesis of stochastic processes structuring the variance, like serial founder effects or isolation by distance, it is expected that Q_{ST} estimates will increase proportionally to the distance between groups, and therefore will be linearly correlated with them. For this study, I calculated the Q_{ST} estimate between all pairs of series within each of the macro-regions, and then averaged them to generate the mean Q_{ST} between pairs of series for each macro-region. The intention with this calculation is to generate a single value that represents the amount of variance that is the result of differences between pairs of series. For each macro-region, a mean linear geographic distance between pairs of series was calculated, and the mean pairwise Q_{ST} estimates within macro-regions were correlated with the mean geographic distance within macro-regions, to test if the Asian macro-regions depart from the pattern observed in other regions of the planet.

Finally, I did a similar exercise, but now between series of each Asian macro-region and series from each of the other macro-regions. Q_{ST} estimates between pairs of series (each of the series in an Asian region to each of the series from another macro-region) were averaged and the values were compared to the average geographic distance between them. In this case, however, geographic distances avoid crossing large bodies of water by going through specific geographic checkpoints (see Hubbe et al. 2010 for a detailed explanation). The goal with this analysis is to test if the accumulation of variance between populations in Asia can be explained solely based on the overall distances that separate these populations from the other macro-regions of the planet. All the analyses were done in R (R Core Team 2019), with functions written specifically for it and complemented with functions from the package MASS (Venables, Ripley, and Venables 2002).

RESULTS: APPORTIONMENT OF MORPHOLOGICAL VARIANCE IN ASIA

The Q_{ST} estimates calculated for the different combinations of the series are presented in Table 3. In general, the results are in accordance with previous worldwide studies of craniometric variance apportionment in modern humans (Relethford 1994, 2004), and show that the vast majority of the morphological variance in humans is partitioned within-groups, while less than 10% is the result of differences between-groups. However, the Q_{ST} values in this study are not directly comparable to previous analysis, since the heritability assumed here is larger. In other words, as the Q_{ST} estimates presented here represent the minimum Q_{ST} possibly obtained ($h^2=1.0$), they are smaller than previous studies that assumed lower heritabilities (usually around 0.55) and reported values between 10 and 20% (e.g., Relethford 1994, 2004).

The Q_{ST} estimates show that, when series are grouped by continents, the apportionment of variance between-groups is smaller ($Q_{ST}=0.051$) than when series are grouped by macro-regions ($Q_{ST}=0.085$), which suggests that continental divisions cause the morphological variance to overlap more among them (i.e., continents are not groupings that maximize the differences between groups), and as such are not necessarily natural boundaries for morphological patterns within them. The higher Q_{ST} for

Table 3.
Q_{ST} values and their standard error calculated for each continent and macro-region, assuming a heritability of 1.0.

Regions	Q_{ST}	Standard error
All continents	0.051	0.001
All macro-regions	0.085	0.001
Asia	0.087	0.002
Australo-Melanesia	0.084	0.002
Africa	0.053	0.002
America	0.097	0.002
Europe	0.052	0.001
East Asia	0.063	0.003
NE Asia	0.056	0.006
South Asia	0.060	0.002
Australia	0.043	0.004
Melanesia	0.074	0.002
East Africa	0.048	0.004
South Africa	0.036	0.004
North Africa	0.020	0.002
West Africa	0.026	0.002
North America	0.054	0.002
Northern North America	0.091	0.003
South America	0.046	0.004
Mediterranean	0.047	0.002
North Europe	0.038	0.002
Polynesia	0.047	0.003

the macro-regions, on the other hand, shows that this geographic category is more efficient at identifying groupings in the series as they are identifying higher amounts of variance between groups. In other words, these Q_{ST} estimate values support the internal coherence of the macro-regions defined here.

The Q_{ST} values for each of the continents show considerable variation among them, with Asia and the Americas presenting almost twice as much variance between-groups (Q_{ST} = 0.087 and 0.097, respectively) than the continents with the smallest Q_{ST} values (Europe; Q_{ST} = 0.053; and Africa; Q_{ST} = 0.052). This indicates a larger degree of differentiation among series in Asia and the Americas than on Europe and Africa, supporting a larger between-groups differentiation in the former two continents. However, from these analyses alone it is impossible to ascertain if these differences are being driven by the geographic and ecological characteristics of these continents, or if they follow broader global patterns of isolation by distance, since the average distance between series in Asia and America (4252.0 km and 4775.0 km, respectively) are also larger than the average distances between series in Europe (2027.8 km) and Africa (3129.9 km). The results of the Relethford-Blangero analysis for the continental division (Fig. 2; Table 4) supports high levels of morphological variance within Asia, as the observed phenotypic variance in this continent is well above the expected one (defined by the regression line in the plot). The other continents show observed phenotypic variances very close to the expected ones, supporting the unique variance levels of Asian populations on a comparative continental level.

When the series are grouped by the macro-regions, the overall Q_{ST} values in Asia decline considerably (Table 3). This change in pattern indicates that the high observed Q_{ST} values observed in Asia as a whole is a product of differences between the three macro-regions of the continent, and that differences within regions are not of the same magnitude. This interpretation is confirmed by the Relethford-Blangero Analysis (Fig. 3; Table 5), which shows that one of the Asian regions (South Asia) have observed phenotypic variance smaller than the expected ones. Northeast Asia and East Asia still show a higher than expected phenotypic variance, and therefore presents a different pattern of variance structure than the other Asian region. None of the Asian macro-regions, however, show levels of variance as high as South Africa, which appears as a clear outlier in the analysis. While outside of the scope of this chapter, it is noteworthy to point out that this pattern in South Africa is consistent with the evidence of high biological diversity in the African continent in general (Betti et al. 2009; Manica et al. 2007).

The last group of analyses carried out for this study relates to the comparison of the average Q_{ST} to geographic distances among the samples with the purpose of exploring the role that isolation by distance played in structuring the morphological variance within and between macro-regions. Figure 4 shows the association between the average Q_{ST} within macro-regions and the average distance in kilometers within

Fig. 2.
Relationship between observed mean observed morphological variance and distance from the centroid (Rii) for the continents included in the study. The regression line indicates the expected morphological variance based on the distance from centroids (Rii), which indicates which continents show higher or lower than expected observed variances.

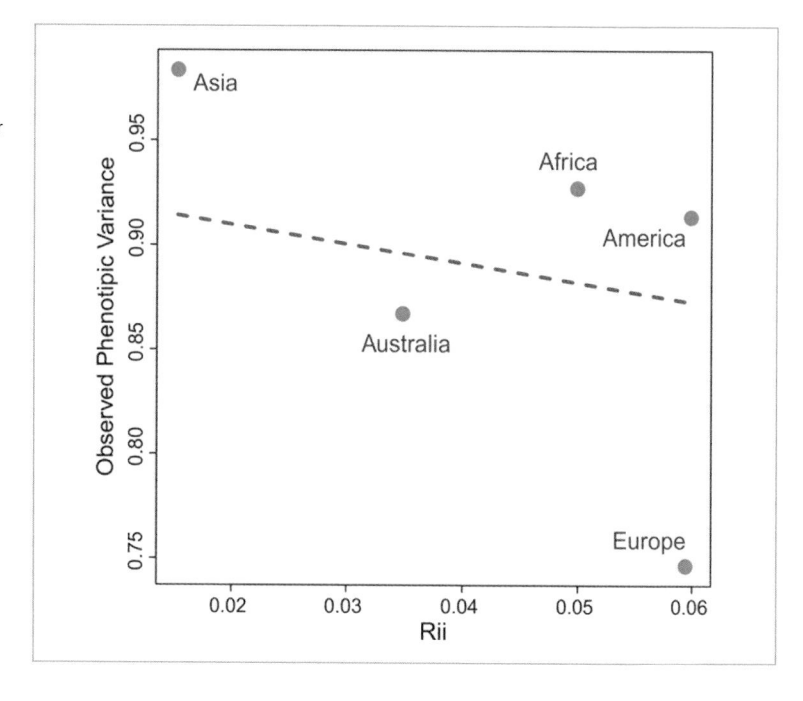

Table 4.
Results of the Relethford-Blangero analysis among continents.
*Rii = genetic distance from centroid (Relethford and Blangero 1990).

	Rii*	Observed Variance	Expected Variance	Residual
Asia	0.015	0.984	0.914	0.069
Africa	0.050	0.927	0.882	0.045
America	0.060	0.914	0.873	0.041
Australia	0.035	0.867	0.896	-0.029
Europe	0.059	0.747	0.873	-0.127

macro-regions. With the exception of Polynesia, all macro-regions show a very strong linear relationship between Q_{ST} and geographic distance (r=0.895, p<0.001; with Polynesia, r=0.572, p=0,027), which indicates a very strong geographic logic to the differentiation of groups within these macro-regions. These results support the idea that, across the planet, regional differences in cranial morphology are regulated by diminishing gene flow between groups as distance between populations increase. The lower relationship of Q_{ST}/km observed in Polynesia is an interesting exception, as it actually lends support to differences being mediated by reduced gene flow. Although distances between series in Polynesia are on average larger than on any of the continents, traveling time between them in the past was not, as Polynesian efficient seafaring facilitated the crossing of larger distances. As such, the effective gene flow between populations was larger than what would be expected if groups were moving across land, since traveling between islands is facilitated by technologi-

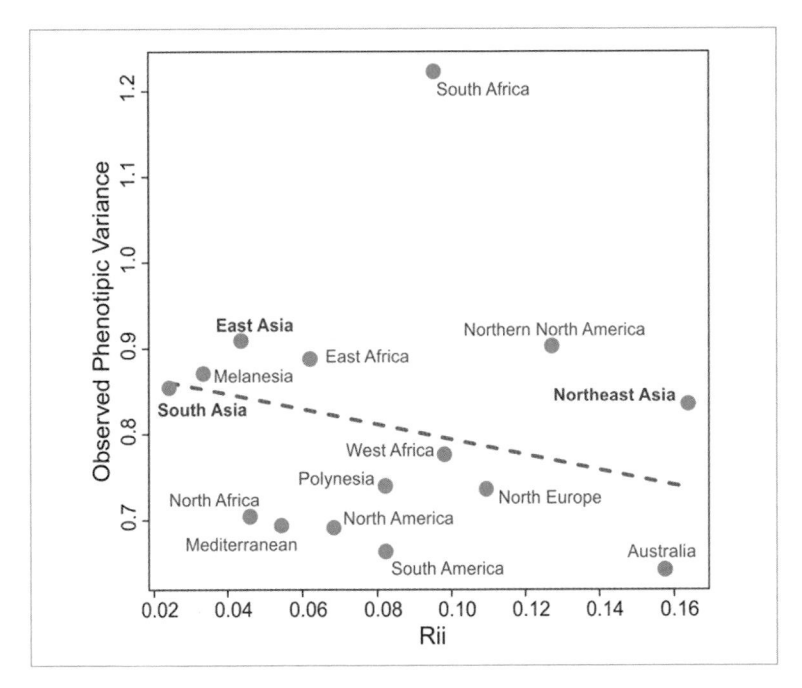

Fig. 3.
Relationship between observed mean morphological variance and distance from the centroid (Rii) for the macro-regions included in the study. The regression line indicates the expected morphological variance based on the distance from centroids (Rii), which indicates which macro-regions show higher or lower than expected observed variances.

Table 5.
Results of the Relethford-Blangero analysis among macro-regions.
*Rii = genetic distance from centroid (Relethford and Blangero 1990).

	Rii*	Observed Variance	Expected Variance	Residual
East Asia	0.044	0.909	0.843	0.066
NE Asia	0.164	0.835	0.737	0.098
South Asia	0.024	0.854	0.861	-0.006
Australia	0.158	0.642	0.743	-0.101
Melanesia	0.033	0.871	0.852	0.018
East Africa	0.062	0.888	0.827	0.061
South Africa	0.095	1.223	0.798	0.425
North Africa	0.046	0.704	0.841	-0.137
West Africa	0.098	0.776	0.795	-0.019
North America	0.068	0.691	0.822	-0.131
Northern North America	0.127	0.903	0.770	0.133
South America	0.082	0.663	0.809	-0.146
Mediterranean	0.054	0.694	0.834	-0.140
North Europe	0.109	0.736	0.785	-0.050
Polynesia	0.082	0.740	0.809	-0.070

Fig. 4.
Relationship between average Q_{ST} and geographic distance within macro-regions. The lighter circle in the lower right portion of the plot represents Polynesia, the only outlier in the 15 macro-regions. The regression line shown is without Polynesia.

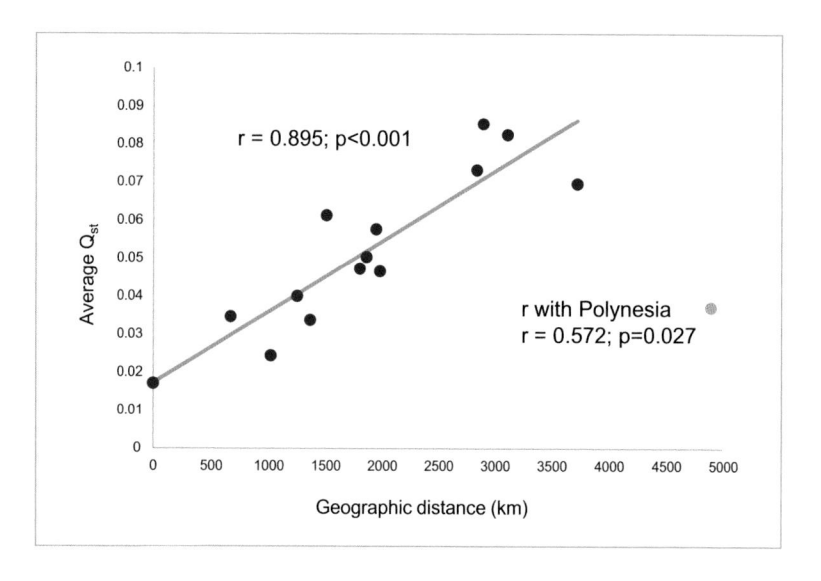

cal innovations. In other words, the Polynesia Q_{ST}/km ratio falls well within the expectation of gene flow regulating the structure of variance between these populations.

This scenario of isolation by distance structure of variance is not sustained, however, when comparisons are done between macro-regions. Figure 5 shows the relationship of average Q_{ST} to average distance in kilometers between each of the Asian macro-regions and the other regions of the planet. As is evident in this plot, a linear isolation by distance is not enough to explain the differences between continents, as none of the Asian macro-regions show a significant correlation between Q_{ST} and geographic distance, when compared to macro-regions outside of Asia. Therefore, isolation by distance is not structuring the morphological variance among macro-regions in Asia. This is somewhat expected, as gene flow will effectively drop to values near zero after enough distance has been accumulated between populations, even if they are connected by other populations in between, so that after that threshold is reached geographic distance will not contribute to the partitioning of variance. For the series included here, this threshold seems to be about five thousand kilometers (Fig. 5), which is the point that separates distances within macro-regions (highly correlated with geography) and distances between macro-region (uncorrelated with geography). On the other hand, if morphological variance continued to be added linearly into the separating populations, like for example through the accumulation of neutral mutations over generations or serial founder-effects, there would still be an increase with distance, even after gene flow stops between populations. The fact that this is not the case (at least in Asia), suggests that morphological variance is being maintained stable over larger geographic spaces, possibly as a result of stabilizing selection.

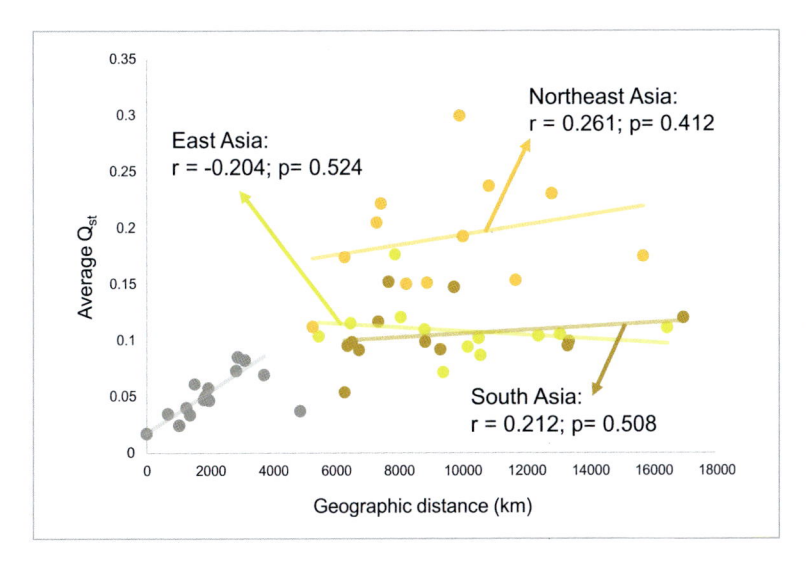

Fig. 5.
Relationship between average Q_{ST} and geographic distance between each of the Asian macro-regions and other regions of the planet. The gray circles and regression line on the left corner of the plot are showing the same data as in Figure 4 (within macro-region distances) as a reference for the scale of differences between analyses.

What is quite remarkable from this analysis is that the average Q_{ST} among Asian macro-regions and the rest of the planet is not the same (ANOVA F = 22.45; p<0.001). While South and East Asian series show similar average Q_{ST} values (Tukey HSD p=0.971) with other macro-regions, Northeast Asia shows higher average values (Tukey HSD p<0.001 for both pairwise comparisons). This suggests that not all Asian macro-regions follow the same pattern of partitioning of variance and that factors unique to Northeast Asian populations differentiate this series from the other two Asian regions, by increasing its morphological distances (i.e., the proportion of variance due the difference between groups) to other regions in the planet. Therefore, while within each of the macro-regions a similar geographic structure is observed across the entire planet (Fig. 4), the variance between macro-regions is being structured by different kind of processes (Fig. 5). These results suggest that the evolutionary forces shaping the morphological variance in the planet change according to the scale of the analysis.

DISCUSSION: THE STRUCTURE OF MORPHOLOGICAL VARIANCE IN ASIA

The analysis of the apportionment of cranial morphological variance across Asia shows the complexities in trying to understand the evolutionary forces that are structuring phenotypic variance over time and space in modern humans. Although previous studies have supported that the morphological differentiation on a global scale tends to follow patterns similar to what would be expected in phenotypic traits evolving neutrally (Betti et al. 2009; Manica et al. 2007; Relethford 1994, 2004; Roseman 2004; von Cramon-Taubadel and Weaver 2009), a more nuanced analysis, as the one presented here, shows that this may be a simplistic view of the processes structuring our species' morphological variance. Acknowledging this complexity is particularly relevant in the context of studies of human dispersion, since they depend on the reconstruction of the ancestral condition of the populations being sampled, and inaccurate assumptions about the forces that promoted the differentiation of the populations over space and time can lead to vastly incorrect reconstructions of the dispersion paths under study.

In the context of this discussion, the results presented here provide a cautionary tale for the general assumptions that are usually informing studies about past modern human migration and dispersion routes. As shown with the analyses presented here, the hypothesis that the forces that structure the morphological variance among modern humans worldwide are relatively constant cannot be easily supported. Indeed, what we observe is that the scale of analysis is extremely relevant in the way by which morphological variance is structured worldwide. While there is a clear pattern of differentiation that follows a geographic structure on a global scale (Betti et al. 2009; Howells 1989; Manica et al. 2007; Relethford 1994), the rate of this differentiation or the degree to which geography correlates with morphological differentiation does not appear to be constant or linear. The results presented here show that the morphological differentiation of the skull does not correlate linearly with geography when Asian macro-regions are compared to other macro-regions of the planet, even though within macro-regions, morphological distances show a strong geographic structure. Therefore, it is possible that the morphological differentiation between larger macro-regions in the planet follows a pattern closer to punctuated equilibrium, with morphological variance being restrained for longer periods of time, before selective pressures are released and a quick influx of morphological variance is observed. Such a pattern of evolution through stasis and quick bursts would still create a global structure of variance that is similar to that resulting from sequential founder-effects and isolation by distance. Moreover, such a model of morphological differentiation fits better with the current evidence of early African morphological patterns being retained among Late Pleistocene populations in Asia (Harvati 2009; Hubbe et al. 2011), Australo-Melanesia (Schillaci 2008) and the Americas (Hubbe et al. 2011). Even though the analyses presented here are not enough to elucidate the mechanism that is structuring the morphological variance within and between macro-regions in Asia and in the planet, they do open the possibility to consider that there are other ways by which morphological variance is being structured in the continent. In other words, they challenge the assumption that uniform processes of differentiation explain the diversification of modern human cranial morphology across the planet or across different spatial scales of analysis.

As such, the analyses presented here suggest that the forces structuring the morphological variance in Asia may follow or depart from global patterns, depending on the scale of analysis. When the morphological variance within macro-regions is analyzed, a strong geographic signal of gene flow mediation is observed in all macro-regions of the planet, including the three macro-regions of Asia. However, there are different responses between macro-regions, as the morphological distances between Asian regions and other macro-regions of the planet show no clear pattern of geographic structure. In other words, the evolutionary forces that might have produced the observed structure in the phenotypic variance between and within modern humans cannot be easily explained by the linear associations of the series over long distances. Moreover, the level of differentiation between regions varies significantly within the Asian macro-regions, with South and East Asian showing similar levels of differentiation to the series from other macro-regions, while Northeast Asia shows significantly higher levels of differentiation. This shows that, on a macro-regional level, there can be significantly different patterns of apportionment of variance, which must be taken into account before creating accurate models of human dispersion in the past.

The analyses of morphological variance apportionment in Asia allows to explore the expectations that were derived from the four major dispersion events identified in Asia. For instance, the results presented here show no evidence of the first stage of dispersion out of Africa in the morphological variance of the series. If the morphology that characterize early populations in the continent and in some isolated groups in SE Asia (Reyes-Centeno et al. 2015) had been a significant source of influx of variance on later populations, or if admixture with other hominin populations was significant in this first stage, it would be expected that the variance apportionment would be inflated within regions and that there would be differences in apportionment between East and South Asia. As the two regions show similar patterns of variance structure among them, the contribution of this early dispersion was probably not significant. This is not to say, however, that the early dispersion or admixture with early hominins did not contribute to the modern morphology through admixture. It means only that it did not contribute to the structuring of variance (i.e., the admixture that happened was distributed across larger portions of the continent over time and did not result in increased difference between populations more closely related to the first dispersion). The distinction is important to be made, as the analyses presented here focus only the apportionment of variance and not on the absolute difference in morphological shape between populations. Evidently, this conclusion is limited to the populations that were included in this study, and the inclusion of more populations that retained the contribution of early morphological dispersion event and admixture in Asia could change this conclusion.

The second large stage of dispersion seems to be strongly associated with the morphological structure in Asia, given the strong geographic structure that exists within the macro-regions. As suggested before, the result for the analyses between regions support a pattern of punctuated equilibrium for the structure of variance in Asia, which can be achieved either through a fast rate of expansion across Asia, followed by isolation by distance models of differentiation after regions have been initially occupied, or by evolutionary constrains (i.e., stabilizing selection) to the accumulation of morphological variance. Any of these processes, as well as the different combinations of them, would result in morphological variances not structured on larger continental scales, while keeping a strong geographic structure within regions.

While it is evident that the expansion across Asia has a strong impact on the morphological variance in the whole continent, other factors play an important role as well. Such is the case for the adaptations seen in populations who colonized colder climates, as described in the third large stage of dispersion in Asia. As shown by a series of previous studies, adaptation to cold environments, mostly as a response to optimizing heat loss in the head (Harvati and Weaver 2006; Hubbe et al. 2009) has promoted significant changes in the morphological pattern of Asian and European populations living in high latitudes (Evteev et al. 2014; Hubbe et al. 2009; Evteev, this volume). These changes probably were complemented by adaptations that resulted from changes in diet and cultural buffering that permitted the occupation of the higher latitude in the continent. While previous studies have not explored this possibility in Asia per se, cultural changes in diet and behavior can have important impact in the overall cranial shape of populations (Galland et al. 2016; González-José et al. 2005).This adaptative response to cold climates shows an important impact in the structure of morphological variance of Northeast Asian populations. Northeast Asia shows a high within-group variance (Fig. 3), which is different from the other two Asian macro-regions. However, this variance is still structured geographically among the series within the macro-region, following the pattern seen worldwide. Finally, it differs again from the others in the average distance between this region and other regions of the planet, which suggests that the adaptation to colder environments is responsible for an important portion of the variance structure in the region. Interestingly, the differences promoted by cold adaptation do not erase the other forces structuring morphological variance, acting as an additive component to the variance between groups.

Lastly, the fourth stage of human dispersion across Asia, which is associated with the increased mobility of human groups during the last two thousand years, seems to have had a smaller impact on the overall structure of the morphological variance in the continent. The geographic structure within Asian regions is comparable to other regions of the planet, which is not what would be expected if gene flow had increased significantly between populations separated by longer distances in the continent. While this is not a new find and several morphologic and genetic studies have described the strong spatial structure within the continent (HUGO Pan-Asian SNP Consortium et al. 2009; Reyes-Centeno et al. 2014; Reyes-Centeno et al. 2015), it is still worth discussing why the intense human mobility in Asia (and in the planet as a whole) that characterizes the second half of the Holocene, did not have a more visible impact on the structure of the morphological variance. The present study lacks the data to explore this point in more detail, but it should be a topic of further research, given the potential impact that the more recent events had in reshaping the cultural background of Asian populations.

In conclusion, this chapter highlights the importance of understanding morphological variance from the point of view of the different evolutionary forces shaping the structure of the phenotypic variance between and within populations, as this is a central aspect for estimating ancestral morphological characteristics of human populations, and is fundamental to trace back the most likely dispersion events and migration routes that connect those populations over space and time. As I tried to demonstrate with the analysis of morphological variance across three different macroregions in Asia, an assumption that populations over large regions were differentiated through similar evolutionary forces acting over different scales of space and time is not an accurate approach for reconstructing past dispersion events. On the contrary, it will result in an inaccurate estimate of population movements in the past. As such, a better understanding of the evolutionary forces structuring the morphological variance, derived from studies conducted by combining different datasets (morphological, genetic, linguistic, archaeological), is an important step in refining our ability to better comprehend the ways by which human groups colonized and expanded across the planet.

ACKNOWLEDGMENTS

I want to thank Hugo Reyes-Centeno, Katerina Harvati, and Gerhard Jäger for the invitation to participate on the symposium that inspired this volume. I am eternally grateful to Prof. Tsuheniko Hanihara for making his dataset available for this study.

REFERENCES

Balaresque, P., N. Poulet, S. Cussat-Blanc, P. Gerard, L. Quintana-Murci, E. Heyer, and M. A. Jobling. 2015. Y-chromosome descent clusters and male differential reproductive success: Young lineage expansions dominate Asian pastoral nomadic populations. *European Journal Of Human Genetics* 23: 1413.

Betti, L., F. Balloux, W. Amos, T. Hanihara, and A. Manica. 2009. Distance from Africa, not climate, explains within-population phenotypic diversity in humans. *Proceedings of the Royal Society B: Biological Sciences* 276 (1658): 809–814.

Carlson, D. S., and D. P. Van Gerven. 1979. Diffusion, biological determinism, and biocultural adaptation in the Nubian Corridor. *American Anthropologist* 81 (3): 561–580.

Carson, E. A. 2006. Maximum likelihood estimation of human craniometric heritabilities. *American Journal of Physical Anthropology* 131 (2): 169–180.

Darroch, J. N., and J. E. Mosimann. 1985. Canonical and Principal Components of Shape. *Biometrika* 72 (2): 241–252.

de Azevedo, S., A. Nocera, C. Paschetta, L. Castillo, M. Gonzalez, and R. Gonzalez-Jose. 2011. Evaluating microevolutionary models for the early settlement of the New World: the importance of recurrent gene flow with Asia. *American Journal of Physical Anthropology* 146 (4): 539–552.

Evteev, A., A. L. Cardini, I. Morozova, and P. O'Higgins. 2014. Extreme climate, rather than population history, explains mid-facial morphology of northern asians. *American Journal of Physical Anthropology* 153 (3): 449–462.

Galland, M., D. P. Van Gerven, N. von Cramon-Taubadel, and R. Pinhasi. 2016. 11,000 years of craniofacial and mandibular variation in Lower Nubia. *Scientific Reports* 6: 31040.

González-José, R., F. Ramirez-Rozzi, M. Sardi, N. Martinez-Abadias, M. Hernandez, and H. M. Pucciarelli. 2005. Functional-cranial approach to the influence of economic strategy on skull morphology. *American Journal of Physical Anthropology* 128 (4): 757–771.

Hallgrímsson, B., D. E. Lieberman, W. Liu, A. F. Ford-Hutchinson, and F. R. Jirik. 2007. Epigenetic interactions and the structure of phenotypic variation in the cranium. *Evolution & Development* 9 (1): 76–91.

Hanihara, T. 1996. Comparison of craniofacial features of major human groups. *American Journal of Physical Anthropology* 99(3): 389–412.

Hansen, V. 2012. *The Silk Road: A new history.* Oxford, New York: Oxford University Press.

Harvati, K. 2009. Into Eurasia: A geometric morphometric re-assessment of the Upper Cave (Zhoukoudian) specimens. *Journal of Human Evolution* 57 (6): 751–762.

Harvati, K., and T. D. Weaver. 2006. Human cranial anatomy and the differential preservation of population history and climate signatures. *The Anatomical Record Part A: Discoveries in Molecular, Cellular, and Evolutionary Biology* 288A (12): 1225–1233.

Howells, W. W. 1989. *Skull shapes and the map: Craniometric analyses in the dispersion of modern Homo.* Cambridge: Peabody Museum of Archaeology and Ethnology Distributed by Harvard University.

Hubbe, M., T. Hanihara, and K. Harvati. 2009. Climate signatures in the morphological differentiation of worldwide modern human populations. *Anatomical Record-Advances in Integrative Anatomy and Evolutionary Biology* 292 (11): 1720–1733.

Hubbe, M., K. Harvati, and W. Neves. 2011. Paleoamerican morphology in the context of European and East Asian Late Pleistocene variation: Implications for human dispersion into the New World. *American Journal of Physical Anthropology* 144 (3): 442–453.

Hubbe, M., W. A. Neves, and K. Harvati. 2010. Testing evolutionary and dispersion scenarios for the settlement of the New World. *Plos One* 5(6): e11105.

HUGO Pan-Asian SNP Consortium, M. A. Abdulla, I. Ahmed, A. Assawamakin, J. Bhak, S. Brahmachari, et al. 2009. Mapping human genetic diversity in Asia. *Science* 326 (5959): 1541–1545.

Jacobs, Z., B. Li, M. V. Shunkov, M. B. Kozlikin, N. S. Bolikhovskaya, et al. 2019. Timing of archaic hominin occupation of Denisova Cave in southern Siberia. *Nature* 565 (7741): 594–599.

Klingenberg, C. P. 2014. Studying morphological integration and modularity at multiple levels: Concepts and analysis. *Philosophical transactions of the Royal Society of London. Series B, Biological sciences* 369 (1649): 20130249-20130249.

Lahr, M. M. n. 1996. *The evolution of modern human diversity: A study of cranial variation.* New York: Cambridge University Press.

Lieberman, D. 2011. *The evolution of the human head.* Cambridge: Belknap Press of Harvard University Press.

Manica, A., W. Amos, F. Balloux, and T. Hanihara. 2007. The effect of ancient population bottlenecks on human phenotypic variation. *Nature* 448: 346.

Mellars, P. 2006. Going east: New genetic and archaeological perspectives on the modern human colonization of Eurasia. *Science* 313 (5788): 796.

Mellars, P., K. C. Gori, M. Carr, P. A. Soares, and M. B. Richards. 2013. Genetic and archaeological perspectives on the initial modern human colonization of southern Asia. *Proceedings of the National Academy of Sciences* 110 (26): 10699–10704.

Neves, W. A., and M. Hubbe. 2005. Cranial morphology of early Americans from Lagoa Santa, Brazil: Implications for the settlement of the New World. *Proceedings of the National Academy of Sciences* 102 (51): 18309–18314.

Petraglia, M. D., M. Haslam, D. Q. Fuller, N. Boivin, and C. Clarkson. 2010. Out of Africa: New hypotheses and evidence for the dispersal of Homo sapiens along the Indian Ocean rim. *Annals of Human Biology* 37 (3): 288–311.

Pinhasi, R. O. N., and N. von Cramon-Taubadel. 2012. A craniometric perspective on the transition to agriculture in Europe. *Human Biology* 84 (1): 45–66.

R Core Team. 2019. R: A language and environment for statistical computing. Retrieved from https://www.R-project.org/

Rasmussen, M., X. Guo, Y. Wang, K. E. Lohmueller, S. Rasmussen, et al. 2011. An Aboriginal Australian genome reveals separate human dispersals into Asia. *Science* 334 (6052): 94–98.

Reich, D., N. Patterson, M. Kircher, F. Delfin, M. R. Nandineni, et al. 2011. Denisova admixture and the first modern human dispersals into Southeast Asia and Oceania. *American Journal of Human Genetics* 89 (4): 516–528.

Relethford, J. H. 1994. Craniometric variation among modern human populations. *American Journal of Physical Anthropology* 95 (1): 53–62.

Relethford, J. H. 2004. Global patterns of isolation by distance based on genetic and morphological data. *Human Biology* 76 (4): 499-513.

Relethford, J. H., and J. Blangero. 1990. Detection of differential gene flow from patterns of quantitative variation. *Human Biology* 62 (1): 5–25.

Reyes-Centeno, H., S. Ghirotto, F. Détroit, D. Grimaud-Hervé, G. Barbujani, and K. Harvati. 2014. Genomic and cranial phenotype data support multiple modern human dispersals from Africa and a southern route into Asia. *Proceedings of the National Academy of Sciences* 111 (20): 7248.

Reyes-Centeno, H., M. Hubbe, T. Hanihara, C. Stringer, and K. Harvati. 2015. Testing modern human out-of-Africa dispersal models and implications for modern human origins. *Journal of Human Evolution* 87: 95–106.

Roseman, C. C. 2004. Detecting interregionally diversifying natural selection on modern human cranial form by using matched molecular and morphometric data. *Proceedings of the National Academy of Sciences of the United States of America* 101 (35): 12824–12829.

Schillaci, M. A. 2008. Human cranial diversity and evidence for an ancient lineage of modern humans. *Journal of Human Evolution* 54 (6): 814–826.

Smith, H. F. 2009. Which cranial regions reflect molecular distances reliably in humans? Evidence from three-dimensional morphology. *American Journal of Human Biology* 21 (1): 36–47.

Strauss, A., M. Hubbe, W. A. Neves, D. V. Bernardo, and J. P. Atui. 2015. The cranial morphology of the Botocudo Indians, Brazil. *American Journal of Physical Anthropology* 157 (2): 202–216.

Teixeira, J. C., and A. Cooper. 2019. Using hominin introgression to trace modern human dispersals. *Proceedings of the National Academy of Sciences* 116 (31): 15327.

Valentin, F., F. Détroit, M. J. T. Spriggs, and S. Bedford. 2016. Early Lapita skeletons from Vanuatu show Polynesian craniofacial shape: Implications for remote oceanic settlement and Lapita origins. *Proceedings of the National Academy of Sciences* 113 (2): 292.

Venables, W. N., B. D. Ripley, and W. N. Venables. 2002. *Modern applied statistics with S*. 4th edition. New York: Springer.

von Cramon-Taubadel, N. 2009. Congruence of individual cranial bone morphology and neutral molecular affinity patterns in modern humans. *American Journal of Physical Anthropology* 140 (2): 205–215.

von Cramon-Taubadel, N. 2011. Global human mandibular variation reflects differences in agricultural and hunter-gatherer subsistence strategies. *Proceedings of the National Academy of Sciences* 108 (49): 19546–19551.

von Cramon-Taubadel, N. 2014. Evolutionary insights into global patterns of human cranial diversity: Population history, climatic and dietary effects. *Journal of Anthropological Sciences* 92: 43–77.

von Cramon-Taubadel, N., A. Strauss, and M. Hubbe. 2017. Evolutionary population history of early Paleoamerican cranial morphology. *Science advances* 3: e1602289.

von Cramon-Taubadel, N., and T. D. Weaver. 2009. Insights from a quantitative genetic approach to human morphological evolution. *Evolutionary Anthropology* 18 (6): 237–240.

Zerjal, T., Y. Xue, G. Bertorelle, R. S. Wells, W. Bao, et al. 2003. The genetic legacy of the Mongols. American *Journal of Human Genetics* 72 (3): 717–721.

CHAPTER 8

Associations between human genetic and craniometric differentiation across North Eurasia: The role of geographic scale

Andrej Evteev[1], Patrícia Santos[2], Alexandra Grosheva[3], Hugo Reyes-Centeno[4,5], Silvia Ghirotto[6]

Abstract

This study sets out to consider the influence of geographical scale on the association between molecular genetic differentiation and craniometric phenotypic differentiation in recent human populations. We employ interpopulation distance measurements for three different anatomical regions of the skull and for three different systems of genetic markers in 30 Eurasian populations. Our original dataset comprises 703 male skulls measured for 21 mid-facial, 15 neurocranial and 6 mandibular measurements, in all cases assessing Mahalanobis distances between populations. Published genetic data of more than 2,000 individuals were summarized by between-population F_{ST} based on allele frequencies of autosomal single nucleotide polymorphisms (SNPs), as well as Cavalli-Sforza distances based on the frequencies of 19 Y-chromosome and 29 mtDNA haplogroups. For different geographical scales of analysis, we used Mantel tests to assess the association of craniometric and genetic inter-population distances for the different cranial regions and genetic markers. Our results show that the level of association between craniometric and genetic distances depends on the part of the skull quantified and on the set of variables employed. In our dataset, this association is much stronger for the mid-face than for the cranial vault. Furthermore, the Mantel test correlation coefficients for the broadest, intercontinental level of analysis are moderate to high, and some are among the highest published so far. They are consistently lower at smaller geographic levels of comparison. Autosomal SNP

---

[1] Anuchin Research Institute and Museum of Anthropology, Lomonosov Moscow State University, Russia.
[2] Department of Life Sciences and Biotechnology, University of Ferrara, Italy.
[3] Vavilov Institute of General Genetics, Russian Academy of Sciences, Russia.
[4] DFG Center for Advanced Studies "Words, Bones, Genes, Tools," University of Tübingen, Germany.
[5] Department of Anthropology and William S. Webb Museum of Anthropology, University of Kentucky, USA.
[6] Department of Mathematics and Computer Science, University of Ferrara, Italy.

© 2021, Kerns Verlag / https://doi.org/10.51315/9783935751377.008
Cite this article: Evteev, A., P. Santos, A. Grosheva, H. Reyes-Centeno, and S. Ghirotto. 2021. Associations between human genetic and craniometric differentiation across North Eurasia: The role of geographic scale. In *Ancient Connections in Eurasia*, ed. by H. Reyes-Centeno and K. Harvati, pp. 157-192. Tübingen: Kerns Verlag. ISBN: 978-3-935751-37-7.

distances exhibit the strongest associations with cranial morphology for almost all anatomical regions and at all geographical levels. Our results are evaluated against the background of previous studies assessing the correlation between craniometric, genetic, and geographic distances, drawing attention to the need for investing much more effort in studying factors affecting the association between genetic and craniometric distances at regional and local geographical levels.

INTRODUCTION

During the last two decades, molecular genetic data have been extensively used to validate the role of craniometric variables as a reliable source of information for reconstructing human population history (Roseman 2004; Harvati and Weaver 2006a; von Cramon-Taubadel 2009a; Evteev and Movsesian 2016; Reyes-Centeno et al. 2017, among others). This endeavor follows a decades-long debate on the evolutionary mechanisms affecting cranial form, particularly the degree of environmental and hereditary effects on cranial dimensions (Boas 1912, 1940; Sparks and Jantz 2002; Relethford 2004b, among others). On theoretical grounds, both biological data sources are expected to reflect the natural history of human populations (Relethford and Harpending 1994). For modern humans, this has been empirically supported by the observation that both molecular genetic diversity and cranial phenotypic diversity within modern human populations decreases from Africa, the continental geographic region of origin in the deep past (e.g., Relethford 2004a; reviewed in Reyes-Centeno 2016). In addition, genetic and cranial diversity between populations, i.e., biological distance, increases as a function of geographical separation between populations, i.e., geographical distance (Ramachandran 2005; Betti et. al. 2010). Methodologically, the approach for validating the utility of craniometric variables is therefore often based on the use of genetic distance data as a "gold standard," where it is employed as a benchmark for establishing the degree of biological variation between human populations. In general, it is thought that the higher the degree of association between craniometric and genetic distances is, the better the former reflects the biological relationship between human populations. In this chapter, we concentrate on this methodological approach for validating the utility of craniometric variables in reconstructing the human past.

Early heuristic studies comparing genetic and craniometric distances at the global level were optimistic in reporting moderate to high correlations between the two types of data (Relethford 2004a; Roseman 2004; Gonzalez-Jose et al. 2004; Harvati and Weaver 2006a, 2006b), which comprised both linear and three-dimensional morphometric variables as well as functional and non-functional regions of the genome. However, a bulk of subsequent studies has shown that the strength of the association between craniometric and genetic distances depends on a number of factors, including the part of the skull studied, morphometric technique applied, system of genetic markers employed and, importantly, on the geographical scale of comparison, i.e., global (intercontinental), continental, or local (Harvati and Weaver 2006a, b; Smith et al. 2007; Smith 2009; von Cramon-Taubadel 2009a, 2009b, 2011a, 2011b, 2016; Ricaut et al. 2010; Reyes-Centeno et al. 2014, 2015, 2017; Herrera et al. 2014; von Cramon-Taubadel and Lycett 2014; Smith et al. 2016; Evteev and Movsesian 2016; Evteev et al. 2017; Moiseyev, de la Fuente 2017). For example, whereas some studies select parts of the skull with respect to hypotheses on skeletal integration and modularity (e.g., von Cramon-Taubadel 2011), others sample skull regions with respect to hypotheses on their phylogenetic utility (e.g., Harvati and Weaver 2006a, 2006b; Smith 2009; von Cramon-Taubadel 2009). In fact, because of distinct research designs, a wide range of statistical association between phenotypic and genetic variation has also been demonstrated in earlier works using anthropometric traits and genetic markers (Hiernaux 1956; Sanghvi 1956; Friedlander et al. 1971; Rothhammer and Spielman 1972; Spielman 1973; Neel et al. 1974; see also Relethford and Lees 1982 and Jorde 1980 for excellent reviews of previous literature). The relative effect of the study design factors listed above remains to be systematically tested since the great majority of recent studies were carried out at the global level and most of these employ only one system of molecular genetic markers.

The question of which anatomical region of the cranium reflects population history best has been a matter of great interest. It is generally agreed that a) the mandible displays the lowest correlations and, within the cranium, b) the temporal bone performs slightly better, probably due to its mostly chondrocranial embryonic origin (Harvati and Weaver 2006a; Nicholson and Harvati 2006; Smith 2009; Reyes-Centeno et al. 2017). However, results of different studies are not directly comparable due to the lack of uniformity in their research designs. It is also unclear if the trends observed at the global (intercontinental) level are applicable to a lower level of population differentiation, i.e., continental or sub-continental. This study therefore sets out to explore to what degree (i) the selection of craniometric and genomic markers, on the one hand, and (ii) the geographic scale of analysis, on the other, affect the association between genetic differentiation and craniometric interpopulation distances across North Eurasia. We employ measurements of three anatomical regions of the skull and three systems of genetic markers, as well as geographic distances across three geographic analytical scales. Importantly, the vast geographical space of North Eurasia has been underrepresented in previous craniometric-genetic association studies and the present work is thus intended to fill this gap. The results obtained for the study's dataset are evaluated against a background of a large number of correlation coefficients between craniometric, genetic, and geographic distances in the literature. Finally, we discuss the implications of our study in the context of a broader debate on the evolution of cranial form in modern humans.

MATERIALS AND METHODS

The cranial sample employed in this study consists of 703 male skulls from 30 Eurasian populations (Table 1) measured by one of the authors (AE; see Evteev et al. 2014 for the intraobserver error test) using a set of standard calipers according to a protocol including 21 mid-facial, 15 neurocranial and 6 mandibular measurements (see Appendix, Table S1, at the end of this chapter). Missing data were imputed by mean substitution with respect to each population, except in cases where an entire cranial region was missing. For example, mandibles were present only in some of the samples (see Table 1 for details of sample sizes). Raw cranial measurements used in this study are available on request. Matrices of Mahalanobis distances were calculated for every set of groups/variables separately for every particular analysis. Each analysis was conducted once using the raw craniometric variables and again using size-standardized variables, calculated by dividing each measurement by the geometric mean of all measurements, per individual (Darroch and Mosimann 1985).

Cranial samples were matched with molecular genetic data based on ethno-linguistic affinities and geographical origin. In total, we analyzed genetic data for three types of loci: autosomal single nucleotide polymorphisms (SNPs), mitochondrial DNA (mtDNA), and Y-chromosome DNA. First, the autosomal data were collected from two SNP chip array sources for published data of 26 Eurasian populations, comprising 1450 individuals in total (Tables S2, S3, Appendix). Data for the Adygeans were used for the Shapsugi cranial sample. Genetic variation between populations for SNP allele frequencies was calculated by between-population F_{ST} (Weir and Cockerham 1984) using the software 4P (Benazzo et al. 2015). Inter-population F_{ST} measures have been found to be highly correlated with Mahalanobis distances for craniometric variables (Reyes-Centeno et al. 2017). Second, we used haplogroup frequencies for the Y-chromosome and mtDNA data (Evteev et al. 2017), compiled from numerous studies (Table S4, Appendix) carried out using very divergent methodological approaches and styles of publishing of raw material (full mitochondrial genomes, or HVSI and II sequences, or frequencies of haplogroups). Thus, the only possible way of compiling these into a single dataset was to employ haplogroup frequencies. The Y-chromosome data for the Ulchi include samples of the Nanai. In total, we used frequencies of 19 Y-chromosome haplogroups and 29 mtDNA haplogroups to calculate Cavalli-Sforza distances (Cavalli-Sforza and Edwards 1967) for each system of genetic markers. In all cases, we compiled distance matrices representing inter-population genetic variation for the different genetic markers (Tables S6–S9, Supplementry online material). Table 1 lists details of the availability of different types of genetic markers for different cranial population samples. In most cases, all three molecular data types could be compiled.

Geographic distances between centroids of origin of all populations (Tables S10–S11, Supplementary online material) were calculated as great-circle distances (using the Haversine formula in: http://www.movable-type.co.uk/scripts/latlong.html), ignoring possible water and mountain barriers. In order to compare our results to previous studies, the strength of association between the matrices of craniometric, genetic, and geographic distances was assessed using Mantel tests (Smouse and Sokal 1986). These were carried out in PAST (Hammer et al. 2011) by setting the similarity measure as "User distance" universally for all matrices and testing for significance via 9999 permutations of the matrix values. Results are reported as Pearson r correlation values with one-tailed p significance values. In addition, we employed Dow-Cheverud tests (Dow and Cheverud 1985) to assess which of the subsets of the variables was statistically more associated with genomic markers in comparison to other cranial subsets. Dow-Cheverud tests were conducted in R using a script coded by M. W. Grabowski and C. C. Roseman.

In order to test the role of the geographical level of comparison (see Fig. 1), we used three areal divisions: intercontinental, continental, and local. First, the main dataset (30 populations, intercontinental level) was divided into two subsets: "West Eurasia" (17 populations, continental level) and "East Eurasia" (12 populations, continental level). These two are based on the separation of Europe and Asia as distinct continents and corresponds with analytical sub-divisions of previous morphological and genetic studies (e.g., Howells 1989; Hanihara 2000; Reich et al. 2012; Fu et al. 2013). As the origin and degree of admixture in the Saami is not completely clear (see Tambets et al. 2004), this population was not included in any of the two subsets. The latter were further divided into

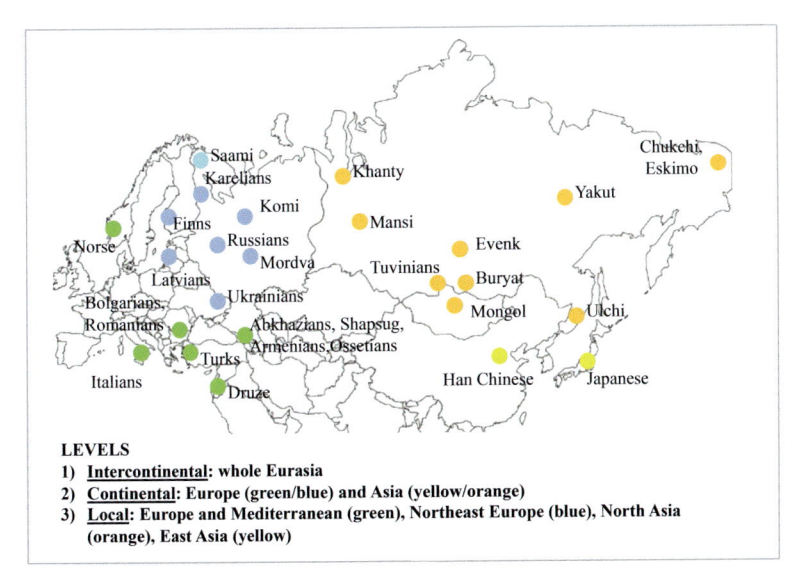

Fig. 1.
Geographic locations of the studied groups.

LEVELS
1) <u>Intercontinental</u>: whole Eurasia
2) <u>Continental</u>: Europe (green/blue) and Asia (yellow/orange)
3) <u>Local</u>: Europe and Mediterranean (green), Northeast Europe (blue), North Asia (orange), East Asia (yellow)

Sample	Genetic data availability[1]	Sample size (facial skeleton)	Sample size (neurocranium)	Sample size (mandible)	Collection[2]
Abkhazian	abc	15	15	0	RIMA
Armenian	abc	26	26	0	RIMA
Bulgarian	abc	15	14	7	RIMA
Buryat	abc	19	19	17	MAE
Chukchi	ac	31	30	14	RIMA
Druze	abc	21	20	0	MAE
Eskimo (Siberian Yupik)	abc	20	19	16	RIMA
Evenk	bc	15	15	8	RIMA / MAE
Finn	abc	21	21	11	RIMA /MAE
Han	abc	20	19	0	RIMA /NHM
Italian	abc	18	18	0	RIMA
Japanese	abc	26	26	23	RIMA / MDH
Karelian	abc	51	43	36	MAE
Khanty	bc	21	21	11	RIMA
Komi	abc	28	28	27	MAE
Latvian	bc	21	20	14	RIMA
Mansi	abc	16	16	0	RIMA
Mongol	abc	18	18	0	MAE
Mordovian	abc	28	25	25	RIMA
Norse	abc	18	16	8	NHM
Ossetian	abc	26	26	0	RIMA
Romanian	abc	32	22	32	FRI
Russian	abc	64	64	60	RIMA
Saami	bc	27	27	24	MAE
Shapsug	abc	15	15	0	RIMA
Turk	abc	11	11	9	RIMA
Tuvinian	abc	26	26	0	RIMA
Ukrainian	abc	12	12	0	RIMA
Ulchi	abc	22	21	17	RIMA / MAE
Yakut	abc	20	20	15	RIMA / MAE
Total number of samples		30	30	19	
Total number of individuals		703	673-683	374	

[1] a: autosomal SNP data is present, b: Y-chromosome data is present, c: mtDNA data is present
[2] Abbreviations: RIMA – Research Institute and Museum of Anthropology, Moscow State University; MAE - Museum of Anthropology and Ethnography (the Kunstkamera), Saint-Petersburg; NHM - Natural History Museum, London; FRI - Francisc Rainer Institute, Bucharest; MDH - Musée de l'Homme, Paris.

Table 1.
Cranial sample populations, genomic data matching, sample sizes, provenance, and geography.

Sub-continent/Region
West Eurasia/ Europe-Mediterranean
West Eurasia/Europe-Mediterranean
West Eurasia/Europe-Mediterranean
East Eurasia/North Asia
East Eurasia/North Asia
West Eurasia/Europe-Mediterranean
East Eurasia/North Asia
East Eurasia/North Asia
West Eurasia/Northeast Europe
East Eurasia/East Asia
West Eurasia/Europe-Mediterranean
East Eurasia/East Asia
West Eurasia/ Northeast Europe
East Eurasia/North Asia
West Eurasia/Northeast Europe
West Eurasia/Northeast Europe
East Eurasia/North Asia
East Eurasia/North Asia
West Eurasia/Northeast Europe
West Eurasia/Europe-Mediterranean
West Eurasia/Europe-Mediterranean
West Eurasia/Europe-Mediterranean
West Eurasia/Northeast Europe
West Eurasia/Europe-Mediterranean
West Eurasia/Europe-Mediterranean
East Eurasia/North Asia
West Eurasia/Northeast Europe
East Eurasia/North Asia
East Eurasia/North Asia

two parts each according to the previously shown ecogeographic differences across those geographical levels (Evteev et al. 2014, 2017). The West Eurasian set was divided into a "Europe-Mediterranean" subset (10 populations, local level), including populations from South and West Europe, the Mediterranean, and the Caucasus, as well as into a "Northeast Europe" (7 populations, local level) subset. The East Eurasian set was divided into "North Asia" (10 populations, local level) and "East Asia" (2 populations, local level). The justification for such a division was the climate-morphology associations demonstrated in previous studies (Evteev et al. 2014, 2017) and separating the populations into presumably cold-adapted and non-cold-adapted. For instance, the Norse were included in the "Europe-Mediterranean" subset because, despite formally representing North Europe, they did not display a particularly strong climatic signal in their craniofacial morphology. Since only two populations were included in the "East Asia" subset, this level was not analyzed further. In total, 6 geographical scales and 3 hierarchical geographic levels of population differentiation were considered. The Karelian sample was excluded from the Mantel tests involving mtDNA matrices in the "West Eurasia" and "North Europe" datasets since it was identified as a genetic outlier. In most analyses, samples genotyped through the Affymetrix array (referred as Affymetrix dataset hereafter) comprising 50786 autosomal SNPs were employed. However, analysis of the Northeast Europe level employed data from the Illumina array (Illumina dataset hereafter), which comprised 114109 markers from 6 populations. The inclusion of both the Affymetrix and Illumina datasets for this particular level increased the number of population samples available for analysis (Tables S2, S3, Appendix). The two datasets provide very similar patterns of interpopulation distances (Mantel Pearson correlation: $r = 0.98$, two-tailed $p = 0.04$ after 1000 permutations) and were thus used interchangeably.

In order to test how the level of genetic or morphological differentiation within a geographical scale affects our Mantel test results, we evaluated the association between the mean Mantel correlation coefficients (Table 1) within a region and either mean autosomal F_{ST} or mean Mahalanobis distances (mid-face) within a region, quantified with the Spearman rank correlation coefficient (r).

In order to compare our results to previous analyses, a database containing about two hundred correlation coefficients between craniometric and anthropometric, as well as genetic distances in humans and non-human primates, published to date was compiled (Table S12, Supplementary online material). Ninety-six of these coefficients, calculated for worldwide (or intercontinental) modern human cranial samples using Mantel tests (Gonzalez-Jose et al. 2004; Roseman 2004; Harvati and Weaver 2006a, 2006b; Smith et al. 2007; Smith 2009; von Cramon-Taubadel 2009a, 2009b, 2011a, 2011b, 2016; Reyes-Centeno et al. 2017; Smith et al. 2016), were selected to be employed as source of comparison for our results.

A similar database containing more than eighty correlation coefficients between craniometric (anthropometric) and geographic distances, as well as latitude and longitude, in humans was also compiled (Table S13, Supplementary online material). Twenty eight of these coefficients were calculated for worldwide (or intercontinental) modern human cranial samples and geographic distances using Mantel tests (Relethford 2004b; Gonzalez-Jose et al. 2001; Smith et al. 2007; Hubbe et al. 2009, 2010, 2011; Betti et al. 2010; von Cramon-Taubadel 2011b, 2016; Noback and Harvati 2015; Reyes-Centeno et al. 2015) were selected to be compared with our results alongside with six coefficients (Rothhammer and Silva 1990; Lalueza Fox et al. 1996; Fabra and Demarchi 2011; Maley 2011; Weisensee 2013; Hubbe et al. 2014) for samples at lower geographic levels (continental, inter-continental, or local).

RESULTS

Associations at the intercontinental (North Eurasian) level.

Table 2 presents the results of all matrix correlations calculated for different cranial anatomical regions, genetic markers, and geographic distances across the different analytical scales and using the raw dataset. Results are similar when using size-standardized craniometric variables (Table S5, Appendix). All the coefficients at the intercontinental level are moderate to high (Fig. 2). But the strength of the association is clearly different between different systems of genetic markers and different anatomical regions (see Figs. 2 and 3). Correlations with the autosomal SNP data demonstrate the strongest associations, while they are weakest for the Y-chromosome data. Considering the lower levels of population differentiation, the autosomal SNP matrices are consistently more

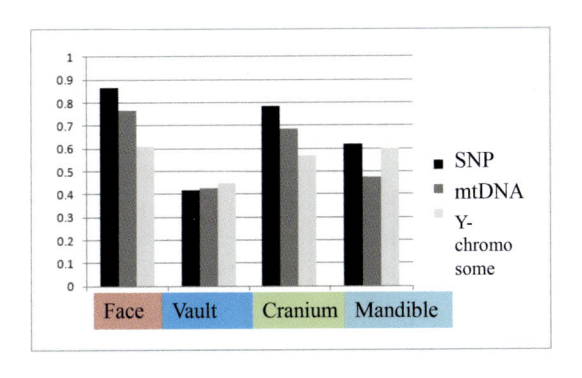

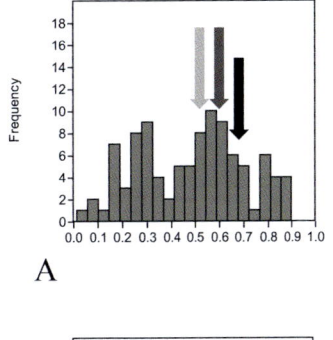

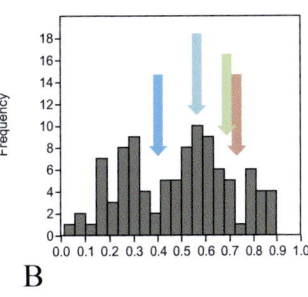

Fig. 2.
Correlations between the matrices of craniometric and genetic distances at the intercontinental analytical level.

Fig. 3.
Mean coefficients obtained in the present study (intercontinental level) against the distribution of 96 coefficients published previously.
A) Different systems of genetic markers (arrows: SNP – black, mtDNA dark grey, Y-chromosome – light grey);
B) Different skull anatomical regions (arrows: face – pink, cranium – green, mandible – light blue, vault – deep blue).

Table 2.
Correlations[1] between the matrices of craniometric, genetic and geographic distances.

Analytical scale	Skull region[2]	SNP	mtDNA	Y-chromosome	Geography
		North Eurasia			
Intercontinental	Mid-face	**0.865*****	**0.767*****	**0.61*****	**0.76*****
	Vault	**0.416*****	**0.428*****	**0.448*****	**0.36*****
	Cranium	**0.783*****	**0.685*****	**0.57*****	**0.66*****
	Mandible	**0.617*****	**0.474*****	**0.6*****	**0.58*****
	mean	0.67	0.589	0.557	0.59
		East Eurasia			
	Mid-face	**0.426***	0.25	**0.5****	**0.39***
	Vault	0.08	0.19	**0.34***	**0.36***
	Cranium	**0.36***	**0.3***	**0.52****	**0.44****
	Mandible	-0.03	-0.2	0.55	0.03
Continental	mean	0.209	0.135	0.478	0.305
		West Eurasia			
	Mid-face	**0.54*****	0.17	**0.39*****	**0.27****
	Vault	**0.16***	0.12	0.02	0.21
	Cranium	**0.29***	0.15	0.19	**0.27***
	Mandible	**0.69****	0.003	**0.47***	**0.49*****
	mean	0.42	0.111	0.268	0.31
		North Asia			
	Mid-face	**0.685****	**0.56****	0.19	**0.5****
	Vault	0.22	0.24	-0.18	**0.44***
	Cranium	**0.55***	**0.45****	-0.005	**0.53*****
	Mandible	-0.17	-0.22	-0.1	-0.21
	mean	0.321	0.258	-0.024	0.315
		Europe-Mediterranean			
Local	Mid-face	**0.35***	0.1	0.21	0.25
	Vault	0.21	0.12	-0.03	0.04
	Cranium	0.2	0.15	0.08	0.11
	mean	0.253	0.123	0.087	0.133
		North Europe			
	Mid-face	0.17	-0.11	0.39	0.27
	Vault	0.2	0.81	0.17	0.18
	Cranium	0.12	0.53	0.26	0.18
	Mandible	-0.31	0.54	0.06	-0.12
	mean	0.045	0.443	0.22	0.128

[1] Values are Pearson r correlation values; bold type indicates statistical significance after 9999 permutations (Mantel test): one-tailed * $p<0.05$; ** $p<0.01$; *** $p<0.001$.
[2] Number of variables in anatomical region: mid-face = 21; vault = 15; cranium (mid-face and vault) = 36; mandible = 6.

associated with the craniometric distance matrices at all levels while the uni-parental markers display various patterns (Table 2; Figs. 5 and 6). The facial skeleton is the part of the cranium exhibiting the highest correlations with genetic distances in all the geographical scales studied (Table 2; Figs. 3, 5 and 6). The mandible displays on average higher coefficients compared to the vault. However, due to small sample sizes, the results for this bone are typically not statistically significant. Within the cranium subsets, the mid-face was significantly more associated with genomic distances than the vault at the highest, intercontintental scale (Dow-Cheverud $r = 0.512$, $p = <0.001$).

Study design factors affecting the association between craniometric and genetic distances at worldwide and intercontinental levels.

Based on our systematic literature review, we found that almost all correlation coefficients published previously (Table S12, Supplementary online material) which can be directly compared with those obtained in the present study were undertaken using worldwide or intercontinental cranial samples, employing geometric morphometric techniques and Mantel tests. Thus, only these geographic levels are considered in this section. However, several systems of genetic markers were used in previous studies and a variety of approaches to sampling of cranial variables were employed. The anatomical regions defined by different authors are numerous and differ with regard to their study design, using, for example, functional anatomical regions or developmental modules. The aggregated categories "braincase," "face," "cranium" and "mandible" are analyzed further. The single anatomical regions for which at least five coefficients were published by at least two authors were considered separately and include the following categories: "basicranium," "neurocranium," "temporal," "vault," "face" and "upper face." The genetic marker systems employed in the literature include: autosomal short-tandem repeats (i.e., STRs or microsatellites), mtDNA from coding regions, "classic" polymorphisms such as blood group or protein markers, and autosomal SNPs.

The mean coefficients for different cranial regions and genetic marker systems in the reviewed literature are presented in Fig. 4a and 4b, respectively, with the inclusion of our original results. An exceptionally high mean was obtained for the vault and, surprisingly, the lowest coefficients were for the neurocranium, which is practically synonymous to the vault. The means of other anatomical regions vary in a relatively narrow range from $r = 0.31$ (mandible) to $r = 0.56$ (cranium). Generalized anatomical regions—cranium, braincase and facial—display slightly higher mean coefficients compared to the others, including the temporal bone and the basicranium, that are argued to contain more phylogenetic information (Lockwood et al. 2004; Harvati and Weaver 2006b; von Cramon-Taubadel 2009a; Reyes-Centeno et al. 2017). The coefficients for the cranium and braincase obtained in the present study for North Eurasia

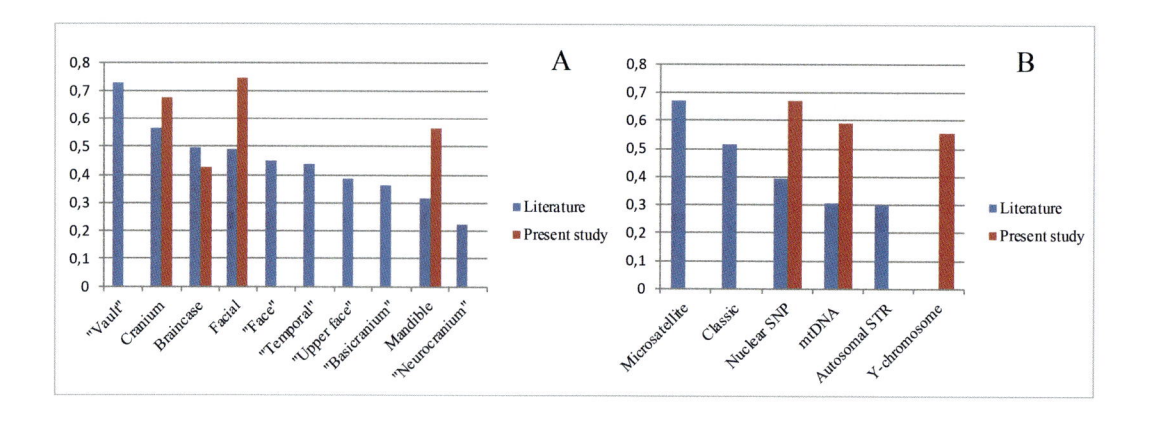

Fig. 4.
Mean correlation coefficients for different cranial anatomical regions and modules (a) and genetic marker systems (b). Reference literature consists of 96 studies. STR I – values published by various authors (Roseman 2004; Harvati and Weaver 2006a, 2006b von Cramon-Taubadel 2009a, 2009b, 2011a); STR II – values from Smith et al. 2007, 2016; Smith 2009.

are very close to the means of respective cranial regions while the coefficients for the facial skeleton and mandible are substantially higher than average (Fig. 4a). An interesting picture is observed for the means of the genetic marker systems (Fig. 4b). For instance, coefficients for the same sets of STRs can vary widely, likely because of a difference in cranial sample composition between different studies: compare STR-I and STR-II (Fig. 4b). The coefficients obtained in the present study for all three systems are about as high as the mean for previous studies using microsatellites. Notably, "classic" (i.e., serological) and uniparenatal markers at the worldwide level do not perform substantially worse than high-throughput nuclear markers.

Geographic distances.

The mean correlation coefficients between geographic and craniometric distances for the four cranial anatomical regions obtained in the present study is almost as high as the means for the autosomal genetic distances (i.e., $r = 0.59$ and $r = 0.67$, respectively; Table 2). It is also substantially higher than the mean ($r = 0.41$) of the 28 coefficients published previously (Table S13, Supplementary online material). The mean calculated across the five lower geographic levels and three cranial anatomical regions (except the mandible), $r = 0.3$, is slightly less than the value for the six analyses of continental, regional, or local levels found in the literature ($r = 0.44$). Importantly, our results show strong associations between geographical distances and the autosomal F_{ST} for some regions (North Eurasia – 0.95; East Eurasia – 0.83; North Asia – 0.82; Northeast Europe – 0.78) but not others (West Eurasia – 0.57; Europe-Mediterranean – 0.66).

Associations at the continental and local levels.

In general, a substantial drop in the strength of association between the cranial metrics, on the one hand, and genetic markers or geography, on the other hand, is observed in the continental data subsets compared to the full North Eurasia (intercontinental) dataset (Table 2; Figs. 5 and 6). The same pattern is observed in the sub-continental levels compared to the continental ones, with the exception of North Asia, where correlation values in some cases increase. As evidenced in the summary of results in Table 2, autosomal SNP distances display the highest correlations with craniometric distances compared to other systems of genetic markers. Likewise, the mid-facial set of measurements consistently displays the highest correlations with genetic distances. Therefore, the two following aspects are considered below in more detail: the association of the mid-facial craniometric distances with various genetic and geographic distances (Fig. 5), and the association of autosomal SNP distances with various craniometric distances (Fig. 6).

While mtDNA distances exhibit higher correlations compared to the Y-chromosome at the intercontinental level, the opposite is true in 4 out

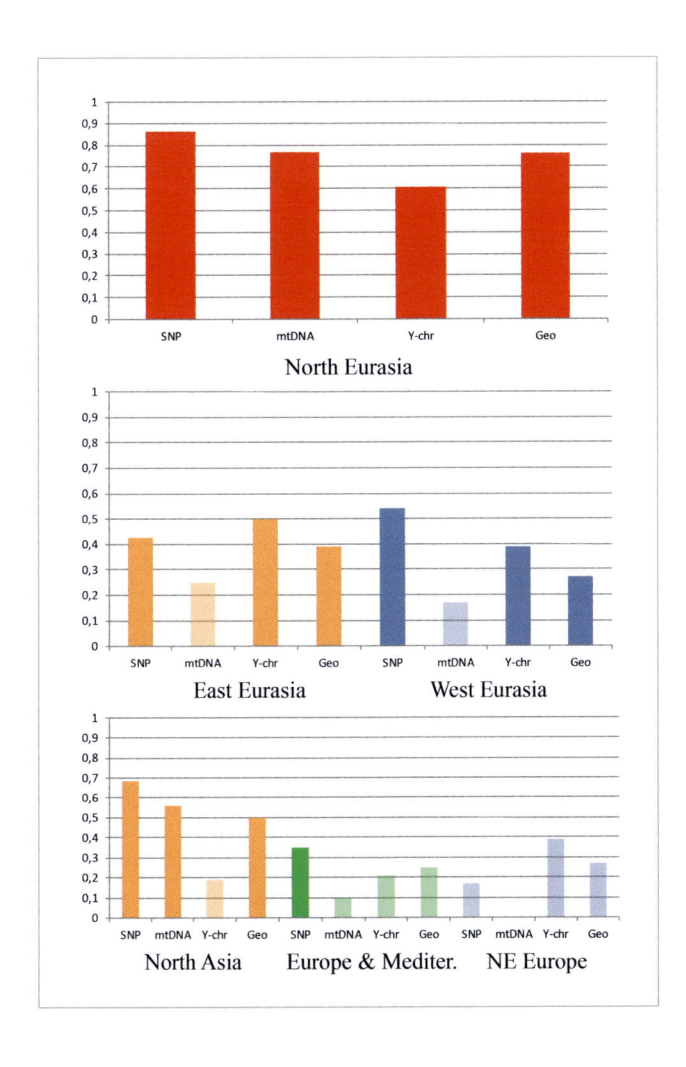

Fig. 5.
The associations between the mid-facial variables, various systems of genetic markers, and geographic distances. Y-axis on all plots corresponds to Pearson r correlation coefficients following Mantel tests. Statistically insignificant coefficients ($p \geq 0.05$) are depicted as transparent bars.

of 5 subsets at lower levels (Fig. 5). Coefficients for the mid-facial metrics (Fig. 6) are typically much higher compared to the vault. The results for the mandible are highly variable, ranging from negative to moderately high positive correlations (though not significant in most cases). The cranium in all geographical subsets displays slightly lower associations compared to the mid-face alone.

The results of the Dow-Cheverud tests for the continental scales show that the difference between the mid-face and vault at this level only applies to West Eurasia (Dow-Cheverud $r = 0.326$, $p = 0.004$) but not to East Eurasia (Dow-Cheverud $r = 0.305$, $p = 0.074$). At the regional scales, the mid-face was significantly more associated with genomic variation only at the North Asian scale (Dow-Cheverud $r = 0.393$, $p = 0.045$).

Fig. 6.
The associations between autosomal SNP markers and various anatomical regions of the skull. Y-axis on all plots corresponds to Pearson r correlation coefficients following Mantel tests. Statistically insignificant coefficients ($p \geq 0.05$) are depicted as transluscent bars.

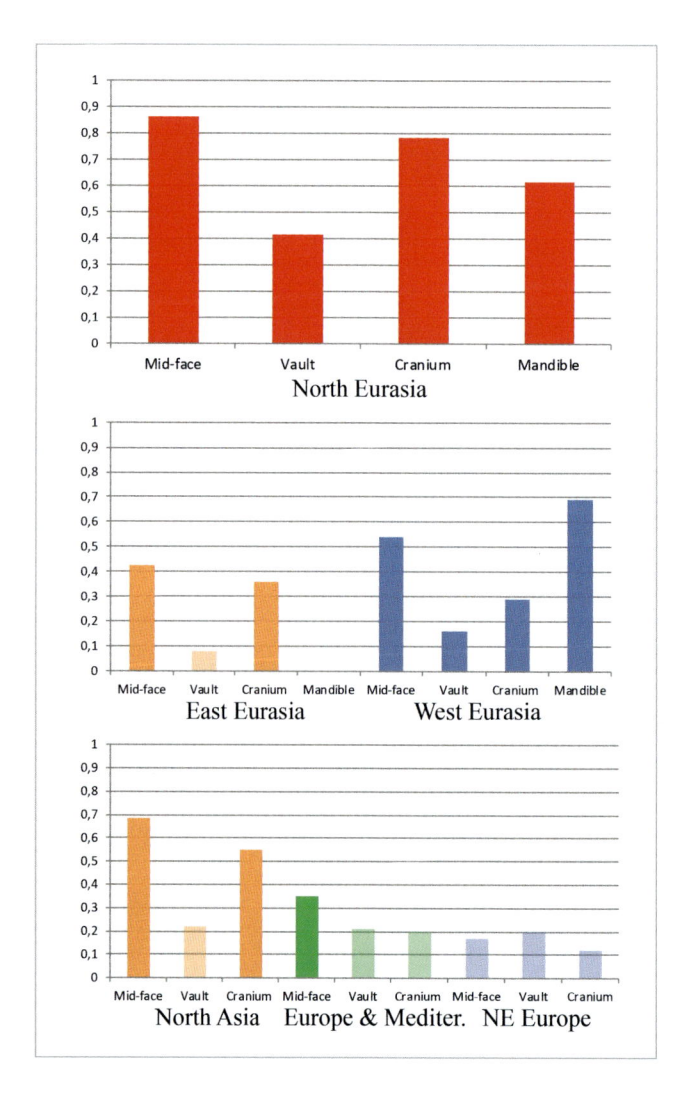

DISCUSSION

The values of the Mantel test coefficients obtained in this study for the intercontinental (North Eurasian) level are moderate to high, which is fully consistent with previous findings at the global level. Some of the coefficients are among the highest published so far. At this level, all the systems of genetic markers employed, as well as geographic distances, display similar results. However, autosomal SNP distances exhibit the strongest associations with cranial morphology in almost all geographical scales and at all levels. This is a predictable result considering that the mode of inheritance of SNPs and cranial morphological traits is similar: numerous loci spread throughout the genome, no sex linkage, relatively low mutation rate, etc. (Lynch 1989; Weaver 2011; Aime et al. 2015). Local variation of the association of cranial morphology with uniparental markers can, in turn, potentially reveal interesting stories about population history when considered against a "background" of autosomal data. For instance, in 4 out of 6 geographical scales Y-chromosome distances are more correlated with cranial morphology than mtDNA distances. We hypothesize that this result might be due, at least in part, to sex-biased migration patterns in different regions of Eurasia. Thus, this and other demographic factors may account for the differential association of craniometric diversity and sex-inherited genetic diversity. In order to further test these inferences using craniometrics, sampling crania of the female sex for the same populations would be necessary. In our opinion, simultaneous use of different systems of genetic markers in craniometric-genetic association studies can provide the most detailed picture of population history, as has been advocated in previous work (Herrera et al. 2014; Evteev and Movsesian 2016). Nevertheless, if only one type of markers is to be used, for instance as a control for population history or phylogeny, the preference, according to our results, should be given to autosomal data.

The results of our study confirm finds of previous works showing that the level of association between craniometric and genetic distances depends on the part of the skull quantified and on the set of variables employed. In general, this association is higher for the mid-face than for the cranial vault in all geographical levels across North Eurasia, as observed by the absolute Mantel test correlation values. The Dow-Cheverud test further showed that this difference is significant at the intercontinental scale, as well as at the West Eurasia and North Asia scales. This is despite the well-established associations of facial shape with climatic conditions in Eurasia (see Evteev et al. 2017 for a review). Importantly, previous work (Howells 1989; Betti et al. 2009) arrived at very similar conclusions for large worldwide datasets. The higher correlations observed for mid-facial traits compared to the vault are not completely unexpected since a number of studies show that the vault is a rather volatile structure which can change rapidly under the influence of a number of factors (Alexeeva 1968; Relethford 2004b). Thus, the mid-facial region is often considered the most important area for ancestry assessment in forensic studies (e.g., Hefner 2009; Scholts et al. 2009). However, the analysis of coefficients published previously demonstrates that the strength of the association is basically identical for the face and braincase, at least at the worldwide level (but see Betti et al. 2009). Thus, the higher coefficients for the mid-facial skeleton might be specific for North Eurasia, suggesting that the cranial features that are most informative of population history or phylogeny might be found in different parts of the skull for different geographical regions.

Turning back to the analysis of literature data, our review sheds light on three primary points. First, it is clear that the cranium in general displays the strongest association with molecular genetics. The same applies to the generalized anatomical regions—face and neurocranium—compared to single bones or smaller units. The morphology of the temporal bone and basicranium do not appear more "phylogenetically relevant" than other modules. According to the same literature analysis, the lowest coefficients among all parts of the skull are typically obtained for the mandible. In the present study, the form of the mandible can display high correlations with genetic distances but these are highly variable across the geographical scales analyzed and are typically not significant. In this regard, it should be noted that sample composition in our study and others is quite different for the mandible compared to other anatomical regions, as the mandible is typically poorly presented in skull collections. Second, our review of the literature also shows that the use of linear measurements instead of landmark-based data employed in most recent works is not inferior in the strength of correlations with genomic data. This is an important observation from the point of view of studying fragmented cranial material since linear measurements can be more easily collected on fragmentary remains and in a cost-effective manner with a standard sliding caliper, in comparison to relatively costly instrumentation required for landmark data acquisition. Third, the results of some previous studies (e.g., Roseman 2004; Roseman et al. 2010; von Cramon-Taubadel 2014) show that the search for phylogenetically informative structures of the skull based on defining a priori modules might not be productive since "…environmental and genetic variation in individual traits are randomly distributed across regions of the cranium rather than being structured by developmental origin or degree of exposure to strain" (Roseman et al. 2010: 1). The use of linear measurements, which often run across different bones and cranial regions, provides an opportunity to apply a potentially more productive "module-free" approach in the search for phylogenetically important variables (see Betti et al. 2009, 2010; Roseman 2004). Likewise, existing (Roseman 2004; Betti et al. 2010; Roseman et al. 2010; Evteev et al. 2020) and novel (e.g., Rathmann and Reyes-Centeno 2020) "module-free" approaches can additionally offer an exhaustive summary of association between anatomical and molecular genetic variation under a framework that conforms to both quantitative genetics and population genetics theory.

With regard to how geography influences the association of cranial and molecular genetic variation, our results show that the correlation of the geographic distance matrices with craniometric ones is about as high as that of autosomal SNP matrices and less variable than the coefficients for uni-parental markers. This holds true for all anatomical regions and levels of comparison. However, the correlations between geographic and genetic matrices differ substantially in different geographical scales studied. Being very high in some levels, where morphological differentiation is strong and geographic distances are large (North Eurasia, North Asia), our results show that it is only moderate in other levels (West Eurasia, Europe-Mediterranean). Thus, geography can only very cautiously be used as a proxy for genetic distances, especially at local levels.

A less optimistic but important finding of this study is the fact that the strength of morphology-genetic association drops consistently and dramatically when moving from the intercontinental scale to lower geographic levels of comparison. In general, this result is expected as a result of the degree of gene flow, where populations close to each other are more likely to meet and exchange genes in comparison to populations far from each other. As such, gene flow is expected to have the immediate effect of homogenizing genetic structure in geographically proximate populations while its consequences for phenotypic variation may be much less pronounced, and this would affect the strength of association between genetic and craniometric data (Reyes-Centeno et al. 2017). Our results can also be explained by numerous confounding factors, the most apparent of which is the imperfect match between cranial and genetic samples. For example, in recent studies on the subject, genetic and morphometric data rarely come from the same individuals and often are not even from the same ethno-linguistic groups. Nevertheless, a number of studies (Hiernaux 1956; Sanghvi 1956; Friedlander et al. 1971; Rothhammer and Spielman 1972; Spielman 1973; Neel et al. 1974; Jorde, 1980; Relethford and Lees 1982) comparing somatometric variables, serological markers, and geographical distances using partially or completely overlapping samples show that even with such a study design, correlations at local levels are not high, and range from 0.17 to 0.55 (Tables S12–S13, Supplementary online material). Another factor is the inherent partial non-neutrality of cranial morphology which is, compared to neutral genetic markers, less affected by stochastic evolutionary factors like genetic drift, mutations, or founder effects. Even if morphological features are not under a direct influence of environmental factors, the need for making a functional structure makes their regulatory and protein-coding genes much more restricted in variation compared to a set of neutral genetic loci (see Lockwood et al. 2004; Weaver 2014). Interestingly, our results show that the strength of morphology-genetic association in North Asia is almost as high as that at the global level. This pattern can be explained by the fact that North Asian populations are extremely diverse in terms of cranial morphology and genetics simultaneously due to the large distances between the populations, their low population sizes, and relative isolation (Debets 1951; Jorde 1980).

Overall, our results provide a pattern that appears to be a general rule: a high level of genetic and/or morphological differentiation in a region leads to an increase of morphology-genetic correlations. Across the six regions studied, mean F_{ST} values based on autosomal markers in a region and Mantel test coefficient values between morphological and genetic distances within regions display a Spearman r correlation of 0.94 (p = 0.003). The same value for mean F_{ST} and mean Mahalanobis distance values is 0.71 (p = 0.14). Thus, the larger craniometric and genetic distances are in a region, the more similar picture of population relationship they tend to provide. As such, our results emphasize the need for investing much more effort in studying factors affecting the association between genetic and craniometric distances at different geographical scales. The strength of this association at worldwide or intercontinental levels is well established and is consistently moderate to high almost irrespectively of the variables used or the part of the skull employed. However, this firm observation can be of little practical value since in the great majority of craniometric studies, the researcher is concerned with interpopulation relationships at much lower geographic scales.

As a whole, our study has implications for the way that craniometric data is utilized in studies that aim to reconstruct the human past. We have shown that geographical scale is an important analytical factor affecting the association between genomic and cranial variation. Because the association between molecular genetic and craniometric datasets is not uniform across different geographical scales in Eurasia, studies that reconstruct population affinities or migration patterns need to explicitly consider the evolutionary forces that might differentially act across geographical space (e.g., see Hubbe, this volume, for a review on the effects of population migration events in Eurasia). Further work is also necessary in understanding how mechanisms in the past have influenced patterns of genomic and cranial variation of recent and present-day populations at different geographical scales.

ACKNOWLEDGMENTS

The authors are grateful to the anonymous reviewer whose comments helped greatly to improve the manuscript. This work was supported by the German Research Foundation (DFG FOR 2237: Project "Words, Bones, Genes, Tools: Tracking Linguistic, Cultural, and Biological Trajectories of the Human Past"). The project was partially supported by the Russian Foundation for Basic Research (grant number 18-56-15001).

REFERENCES

Aime, C., E. Heyer, and F. Austerlitz. 2015. Inference of sex-specific expansion patterns in human populations from Y-chromosome polymorphism. *American Journal of Physical Anthropology* 157: 217–225.

Alexeeva, T. I. 1968. Morpho-Functional population studies in some biogeochemical USSR provinces as viewed in the light of the adaptation problem. In *VIIIth Intern. Congress of Anthropol. and Ethnograph. Sciences*, pp. 1–11. Moscow.

Benazzo, A., A. Panziera, and G. Bertorelle. 2015. 4P: fast computing of population genetics statistics from large DNA polymorphism panels. *Ecology and Evolution* 5: 172–175

Betti, L., F. Balloux, W. Amos, T. Hanihara, and A. Manica. 2009. Distance from Africa, not climate, explains within-population phenotypic diversity in humans. *Proceedings of the Royal Society B: Biological Sciences* 276: 809–814.

Betti, L., F. Balloux, T. Hanihara, and A. Manica. 2010. The relative role of drift and selection in shaping the human skull. *American Journal of Physical Anthropology* 141: 76–82.

Boas, F. 1912. Changes in the bodily form of descendants of immigrants. *American Anthropologist* 14 (3): 530–562.

Boas, F. 1940. Changes in bodily form of descendants of immigrants. *American Anthropologist* 42 (2): 183–189.

Cavalli-Sforza, L. L., and A. W. F. Edwards. 1965. Analysis of human evolution. In *Genetics Today. Proceeding of the XI International Congress of Genetics. Vol. 3*, pp. 923–933. The Hague.

Darroch, J. N., and J. E. Mosimann. 1985. Canonical and principal components of shape. *Biometrika* 72: 241–252.

Debets, G. F. 1951. *Anthropological researches in Kamchatskaya oblast* (Proceedings of the Miklouho-Maclay Institute of Ethnography (New Series). Proceedings of the Northeastern Expedition). Moscow. [in Russian]

Dow, M. M., and J. M. Cheverud. 1985. Comparison of distance matrices in studies of population structure and genetic microdifferentiation: Quadratic assignment. *American Journal of Physical Anthropology* 68 (3): 367–373.

Evteev, A., A. L. Cardini, I. Morozova, and P. O'Higgins. 2014. Extreme climate, rather than population history, explains mid-facial morphology of Northern Asians. *American Journal of Physical Anthropology* 153: 449–462.

Evteev, A. A., and A. A. Movsesian. 2016. Testing the association between human mid-facial morphology and climate using autosomal, mitochondrial, Y chromosomal polymorphisms and cranial non-metrics. *American Journal of Physical Anthropology* 159: 517–522.

Evteev, A. A., A. A. Movsesian, and A. N. Grosheva. 2017. The association between mid-facial morphology and climate in Northeast Europe differs from that in North Asia: Implications for understanding the morphology of Late Pleistocene *Homo sapiens*. *Journal of Human Evolution* 107: 36–48.

Evteev, A. A., P. Santos, S. Ghirotto, and H. Reyes-Centeno. 2020. Associations between genetic differentiation and craniometric interpopulation distances across North Eurasia. *American Journal of Physical Anthropology* 171(S69): 82–82.

Fabra, M., and D. A. Demarchi. 2011. Geographic patterns of craniofacial variation in pre-Hispanic populations from the Southern Cone of South America. *American Journal of Human Biology* 83: 491–507.

Friedlander, J. S., L. A. Sgaramella-Zonta, K. K. Kidd, L. Y. C. Lai, P. Clark, and R. J. Walsh. 1971. Biological divergences in South-Central Bougainville: An analysis of blood polymorphism gene frequencies and anthropometric measurements utilizing tree models, and a comparison of these variables with linguistic, geographic, and migrational "distances." *American Journal of Human Genetics* 23: 253–270.

Fu, Q., M. Meyer, X. Gao, U. Stenzel, H. A. Burbano, J. Kelso, and S. Paabo. 2013. DNA analysis of an early modern human from Tianyuan Cave, China. *Proceedings of the National Academy of Science USA* 110: 2223–2227.

Gonzalez-Jose, R., S. L. Dahinten, M. A. Luis, M. Hernandez, and H. M. Pucciarelli. 2001. Craniometric variation and the settlement of the Americas: Testing hypotheses by means of R-matrix and matrix correlation analyses. *American Journal of Physical Anthropology* 116: 154–165.

Gonzalez-Jose, R., S. Van der Molen, E. Gonzalez-Perez, and M. Hernadez. 2004. Patterns of phenotypic covariation and correlation in modern humans as viewed from morphological integration. *American Journal of Physical Anthropology* 123: 69–77.

Hanihara, T. 2000. Frontal and facial flatness of major human populations. *American Journal of Physical Anthropology* 111:105–134.

Hammer, Ø., D. A. T. Harper, and P. D. Ryan. 2001. PAST: PAlaeontological STatistics. Available at: http://folk.uio.no/ohammer/past/.

Harvati, K., and T. D. Weaver. 2006a. Human cranial anatomy and the differential preservation of population history and climate signatures. *The Anatomical Record A* 288: 1225–1233.

Harvati, K., and T. D. Weaver. 2006b. Reliability of cranial morphology in reconstructing Neanderthal phylogeny. In *Neanderthals Revisited: New Approaches and Perspectives*, ed. by J-J. Hublin, K. Harvati, and T. Harrison, pp. 239–254. Dordrecht: Springer.

Hefner, J. T. 2009. Cranial nonmetric variation and estimating ancestry. *Journal of Forensic Sciences* 54 (5): 985–995.

Herrera, B., T. Hanihara, and K. Godde. 2014. Comparability of multiple data types from the Bering strait region: cranial and dental metrics and nonmetrics, mtDNA, and Y-chromosome DNA. *American Journal of Physical Anthropology* 154: 334–348.

Hiernaux, J. 1956. *Analyse de la variation de characteres physiques humains en une region de l'Afrique centrale: Ruanda-Urundi et Kivu* (Annales du Musee Royal du Congo Belge. Science de l'homme. Ser. in 8o, V. 3). Tervuren.

Howells, W. W. 1989. *Skull shapes and the map.* Cambridge: Peabody Museum Press.

Hubbe, M. 2020. The structure of cranial morphological variance in Asia: Implications for the study of modern human dispersion across the planet. In *Ancient Connections in Eurasia*, ed. by H. Reyes-Centeno and K. Harvati. DFG Center for Advanced Studies Series. Tübingen: Kerns Verlag.

Hubbe, M., T. Hanihara, and K. Harvati. 2009. Climate signatures in the morphological differentiation of worldwide modern human populations. *The Anatomical Record A* 292:1720–1733.

Hubbe, M., W. A. Neves, and K. Harvati. 2010. Testing evolutionary and dispersion scenarios for the settlement of the New World. *PLoS ONE* 5(6):e11105.

Hubbe, M., K. Harvati, and W. Neves. 2011. Paleoamerican morphology in the context of European and East Asian Late Pleistocene variation: Implications for human dispersion into the New World. *American Journal of Physical Anthropology* 144: 442–453.

Hubbe, M., M. Okumura, D. V. Bernardo, and W. A. Neves. 2014. Cranial morphological diversity of Early, Middle, and Late Holocene Brazilian groups: Implications for human dispersion in Brazil. *American Journal of Physical Anthropology* 155: 546–558.

Jorde, L. B. 1980. The genetic structure of subdivided human populations: A review. In *Current Developments in Anthropological Genetics. V.1 Theory and Methods*, ed. by M. H. Crawford and J. H. Mielke, pp. 135–208. New York: Plenum Press.

Lalueza Fox, C., M. Hernandez, and C. Garcia Moro. 1996. Craniometric analysis in groups from Tierra del Fuego/Patagonia and the peopling of the south extreme of the Americas. *Human Evolution* 11: 217–224.

Lockwood, C. A., W. H. Kimbel, and J. M. Lynchd. 2004. Morphometrics and hominoid phylogeny: support for a chimpanzee–human clade and differentiation among great ape species. *Proceedings of the National Academy of Science USA* 101:4356–4360.

Lynch, M. 1989. Phylogenetic hypotheses under the assumption of neutral quantitative-genetic variation. *Evolution* 43 (1):1-17.

Maley, B.C. 2011. Population structure and demographic history of human Arctic populations using quantitative cranial traits. PhD dissertation. Washington University in St. Louis.

Moiseyev, V. G., and C. de la Fuente. 2016. Population history of indigenous peoples of Siberia: Integrating of anthropological and genetic data. *Tomsk State University Journal of History* 5: 158–163.

Neel, J. V., F. Rothhammer, and J. C. Lingoes. 1974. The genetic structure of a tribal population, the Yanomama Indians. X. Agreement between representations of village distances based on different sets of characteristics. *American Journal of Human Genetics* 26: 281–303.

Nicholson, E., and K. Harvati. 2006. Quantitative analysis of human mandibular shape using three-dimensional geometric morphometrics. *American Journal of Physical Anthropology* 131: 368–683.

Noback, M. L., and K. Harvati. 2015. The contribution of subsistence to global human cranial variation. *Journal of Human Evolution* 80: 34–50.

Ramachandran, S., O. Deshpande, C. C. Roseman, N. A. Rosenberg, M. W. Feldman, and L. L. Cavalli-Sforza. 2005. Support from the relationship of genetic and geographic distance in human populations for a serial founder effect originating in Africa. *Proceedings of the National Academy of Sciences USA* 102 (44): 15942–15947.

Reich, D., N. Patterson, D. Campbell, et al. 2012. Reconstructing Native American population history. *Nature* 488: 370–374.

Relethford, J. H., and H. C. Harpending. 1994. Craniometric variation, genetic theory, and modern human origins. *American Journal of Physical Anthropology* 95 (3): 249–270.

Relethford, J. H., and F. C. Lees. 1982. The use of quantitative traits in the study of human population structure. *Yearbook of Physical Anthropology* 25: 113–132.

Relethford, J. H. 2004a. Global patterns of isolation by distance based on genetic and morphological data. *Human Biology* 76 (4): 499–513.

Relethford, J. H. 2004b. Boas and beyond: Migration and craniometric variation. *American Journal of Human Biology* 16: 379–386.

Rathmann, H., and H. Reyes-Centeno. 2020. Testing the utility of dental morphological trait combinations for inferring human neutral genetic variation. *PNAS* 117 (20): 10769–10777.

Reyes-Centeno, H. 2016. Out of Africa and into Asia: Fossil and genetic evidence on modern human origins and dispersals. *Quaternary International* 416: 249–262.

Reyes-Centeno, H., S. Ghirotto, F. Détroit, D. Grimaud-Hervéc, G. Barbujani, and K. Harvati. 2014. Genomic and cranial phenotype data support multiple modern human dispersals from Africa and a southern route into Asia. *Proceedings of the National Academy of Science* USA 111: 7248–7253.

Reyes-Centeno, H., K. Harvati, and G. Jäger. 2016. Tracking modern human population history from linguistic and cranial phenotype. *Scientific Reports* 6 (1): 36645.

Reyes-Centeno, H., M. Hubbe, T. Hanihara, C. Stringer, and K. Harvati. 2015. Testing modern human out-of-Africa dispersal models and implications for modern human origins. *Journal of Human Evolution* 87: 95–106.

Reyes-Centeno, H., S. Ghirotto, and K. Harvati. 2017. Genomic validation of the differential preservation of population history in modern human cranial anatomy. *American Journal of Physical Anthropology* 162: 170–179.

Ricaut, F. X., V. Auriol, N. von Cramon-Taubadel, C. Keyser, P. Murail, B. Ludes, and E. Crubezy. 2010. Comparison between morphological and genetic data to estimate biological relationship: The case of the Egyin Gol necropolis (Mongolia). *American Journal of Physical Anthropology* 143: 355–364.

Roseman, C. C. 2004. Detecting interregionally diversifying natural selection on modern human cranial form by using matched molecular and morphometric data. *Proceedings of the National Academy of Science USA* 101: 12824–12829.

Roseman, C. C., K. E. Willmore, J. Rogers, C. Hildebolt, B. E. Sadler, J. T. Richtsmeier, and J. M. Cheverud. 2010. Genetic and environmental contributions to variation in baboon cranial morphology. *American Journal of Physical Anthropology* 143: 1–12.

Rothhammer, F., and R. S. Spielman. 1972. Anthropometric variation in the Aymará: Genetic, geographic, and topographic contributions. *American Journal of Human Genetics* 24: 371–380.

Rothhammer, F., and C. Silva. 1990. Craniometrical variation among South American prehistoric populations: Climatic, altitudinal, chronological, and geographic contributions. *American Journal of Physical Anthropology* 82: 9–17.

Sanghvi, L. D. 1953. Comparison of genetical and morphological methods for a study of biological difference. *American Journal of Physical Anthropology* 11: 385–404.

Sholts, S. B., P. L. Walker, S. C. Kuzminsky, K. W. P. Miller, and S. K. T. S. Wärmländer. 2011. Identification of group affinity from cross-sectional contours of the human midfacial skeleton using digital morphometrics and 3D laser scanning technology. *Journal of Forensic Science* 56 (2): 333–338.

Smith, H. F., C. E. Terhune, and C. A. Lockwood. 2007. Genetic, geographic, and environmental correlates of human temporal bone variation. *American Journal of Physical Anthropology* 134: 312–322.

Smith, H. F. 2009. Which cranial regions reflect molecular distances reliably in humans? Evidence from three-dimensional morphology. American *Journal of Human Biology* 21: 36–47.

Smith, H. F., B. I. Hulsey, F. L. West (Pack), and G. S. Cabana. 2016. Do biological distances reflect genetic distances? A comparison of craniometric and genetic distances at local and global scales. In *Biological Distance Analysis: Forensic and Bioarchaeological Perspectives*, ed. by M. A. Pilloud and J. T. Hefner, pp. 157–179. Elsevier Science.

Sparks, C. S. and R. L. Jantz. 2002. A reassessment of human cranial plasticity: Boas revisited. *Proceedings of the National Academy of Sciences* 99 (23): 14636–14639.

Spielman, R. S. 1973. Differences among Yanomama Indian villages: Do the patterns of allele frequencies, anthropometrics and map locations correspond? *American Journal of Physical Anthropology* 39: 461–480.

Tambets, K., S. Rootsi, T. Kivisild, et al. 2004. The Western and Eastern roots of the Saami—the story of genetic "outliers" told by mitochondrial DNA and Y chromosomes. *American Journal of Human Genetics* 74: 661–682.

von Cramon-Taubadel, N. 2009a. Congruence of individual cranial bone morphology and neutral molecular affinity patterns in modern humans. *American Journal of Physical Anthropology* 140: 205–215.

von Cramon-Taubadel, N. 2009b. Revisiting the homoiology hypothesis: The impact of phenotypic plasticity on the reconstruction of human population history from craniometric data. *Journal of Human Evolution* 57: 179–190.

von Cramon-Taubadel, N. 2011a. The relative efficacy of functional and developmental cranial modules for reconstructing global human population history. *American Journal of Physical Anthropology* 146: 83–93.

von Cramon-Taubadel, N. 2011b. Global human mandibular variation reflects differences in agricultural and hunter-gatherer subsistence strategies. *Proceedings of the National Academy of Science USA* 108: 19546–19551.

von Cramon-Taubadel, N., J. T. Stock, and R. Pinhasi. 2013. Skull and limb morphology differentially track population history and environmental factors in the transition to agriculture in Europe. *Proceedings of the Royal Society B* 280: 20131337.

von Cramon-Taubadel, N. 2014. Evolutionary insights into global patterns of human cranial diversity: Population history, climatic and dietary effects. *Journal of Anthropological Sciences* 92: 43–77.

von Cramon-Taubadel, N., and S. J. Lycett. 2014. A comparison of catarrhine genetic distances with pelvic and cranial morphology: Implications for determining hominin phylogeny. *Journal of Human Evolution* 77: 179–186.

von Cramon-Taubadel, N. 2016. Population biodistance in global perspective: Assessing the influence of population history and environmental effects on patterns of craniomandibular variation. In *Biological Distance Analysis: Forensic and Bioarchaeological Perspectives*, ed. by M. A. Pilloud and J. T. Hefner, pp. 425–446. Elsevier Science.

Weaver, T. D. 2011. Rates of cranial evolution in Neandertals and Modern Humans. In *Computational Paleontology*, ed. by A. M. T. Elewa, pp. 165–178. Berlin, Heidelberg: Springer Berlin Heidelberg.

Weaver, T. D. 2014. Brief communication: Quantitative and molecular-genetic differentiation in humans and chimpanzees: Implications for the evolutionary processes underlying cranial diversification. *American Journal of Physical Anthropology* 154: 615–620.

Weir, B. S., and C. C. Cockerham. 1984. Estimating F-statistics for the analysis of population structure. *Evolution* 38: 1358–1370.

Weisensee, K. E. 2013. Exploring the relative importance of spatial and environmental variation on the craniometrics of the modern Portuguese. *Human Biology* 85(5): 673–686.

APPENDIX. SUPPLEMENTARY TABLES S1–S5.

Table S1.
List of skull anatomical regions analyzed and craniometric variables.

Cranial region	Code*	Name
mid-face	57/WNB	Simotic chord
mid-face	-/SIS	Simotic subtense
mid-face	56 (2)/-	Total lateral length of the nasalia
mid-face	50/-	Maxillofrontal (interorbital) breadth
mid-face		Nasal height from infranasion
mid-face	54/ NLB	Nasal breadth
mid-face	Premaxilla-4 (Evteev et al. 2017)	Frontal process height
mid-face	Premaxilla-6 (Evteev et al. 2017)	Height of the superior part of the piriform aperture
mid-face	Premaxilla-8 (Evteev et al. 2017)	Zygoorbitale subtense
mid-face	-/SSS	Zygomaxillary subtense
mid-face	45(3)/-	Zygoorbitale chord
mid-face	≈46/ZMB	Zygomaxillary chord
mid-face	Maxilla-3 (Evteev et al. 2017)	Oblique "cheek height"
mid-face	Maxilla-4 (Evteev et al. 2017)	Lateral length of the body of the maxilla
mid-face	Maxilla-5 (Evteev et al. 2017)	Length of the palate from subspinale
mid-face	Maxilla-6 (Evteev et al. 2017)	Internal breadth of the palate
mid-face	Cavity-2 (Evteev et al. 2017)	Inferior anterior breadth of the nasal cavity
mid-face	Cavity-3 (Evteev et al. 2017)	Posterior height of the nasal cavity
mid-face	Cavity-4 (Evteev et al. 2017)	Inferior posterior breadth of the nasal cavity
mid-face	59/-	Morphological height of the choanae
mid-face	≈ 59(1)/-	Choanae breadth

*   Martin 1928 / Howells 1989.
** Nomenclature of the Pearson's Biometrics School.

Description	Source
Distance between the closest points of two nasomaxillary sutures	Martin 1928; Howells 1989
Subtense to simotic chord	Howells 1989
Infranasion – nasomaxillare	Martin 1928; Evteev et al. 2017
Maxillofrontale-maxillofrontale	Martin 1928; Evteev et al. 2017
Infranasion-nariale	Evteev et al. 2017
Alare-alare (nasolaterale-nasolaterale)	Martin 1928; Howells 1989
Maxillofrontale - the point of intersection of the inferior orbital margin and the tangent line to the lower margin of the sulcus lacrimalis	Evteev et al. 2017
Nasomaxillare - conchale	Evteev et al. 2017
Subtense to the chord between left and right zygoorbitale	Evteev et al. 2017
Subtense from tubspinale to the chord between left and right zygomaxillare	Howells 1989
Zygoorbitale - zygoorbitale	Martin 1928
Zygomaxillare- zygomaxillare	Martin 1928
Sum of the distances from the middle of the zygomaxillary suture to zygoorbitale and zygomaxillare	Evteev et al. 2017
Distance from sphenomaxillare superior to the most inferior point of the foramen infraorbitale	Evteev et al. 2017
Subspinale-staphylion	Evteev et al. 2017
Distance between left and right palatomaxillare laterale	Evteev et al. 2017
Maximal distance between the lateral walls of the nasal cavity below crista conchalis immediately after piriform aperture margin but before hiatus maxillaris	Evteev et al. 2017
From the point where the pterygopalatine suture intersects with the margin of the vomer to the most distant point on the floor of the nasal cavity	Evteev et al. 2017
Maximal distance between the lateral walls of the nasal cavity below crista conchalis anterior to choanae but posterior to hiatus maxillaris	Evteev et al. 2017
Staphylion - hormion	Martin 1928
Distance between the points of intersection of the pterygopalatine suture and crista conchalis	Martin 1928; Evteev et al. 2017

cont. ⟶

Table S1. cont.

Cranial region	Code*	Name
neurocranium	1/GOL	Maximum cranial length
neurocranium	8/XCB	Maximum cranial breadth
neurocranium	17/BBH	Basion–bregma height
neurocranium	5/BNL	Cranial base length
neurocranium	9/M9	Minimum frontal breadth
neurocranium	11/AUB	Biauricular breadth
neurocranium	29/FRC	Nasion-bregma chord
neurocranium	Sub. Nβ (Biometrics**)	Frontal subtense
neurocranium	30/PAC	Bregma-lambda chord
neurocranium		Parietal subtense
neurocranium	31/OCC	Lambda-opisthion chord
neurocranium		Occipital subtense
neurocranium	26/M26	Sagittal frontal arc
neurocranium	27/M27	Saggital parietal arc
neurocranium	28/M28	Saggital occipital arc
mandible	71a/-	Minimum width of the ramus
mandible	65/-	Condylar breadth
mandible	66/-	Bigonial breadth
mandible	67/-	Anterior breadth
mandible	69/-	Symphyseal height
mandible	69(3)/-	Corpus width of the mandible

* Martin 1928 / Howells 1989.
** Nomenclature of the Pearson's Biometrics School.

Description	Source
Glabella – opisthocranion	Martin 1928
Eurion-eurion	Martin 1928
Basion–bregma	Martin 1928
Basion-nasion	Martin 1928
Frontotemporale-frontotemporale	Martin 1928
Auriculare - auriculare	Martin 1928
Nasion-bregma	Martin 1928
Subtense to the nasion-bregma chord	Alekseev and Debets 1964
Bregma-lambda	Martin 1928
Subtense to the bregma-lambda chord	Alekseev and Debets 1964
Lambda-opisthion	Martin 1928
Subtense to the lambda-opisthion chord	Alekseev and Debets 1964
Arc from nasion to bregma	Martin 1928
Arc from bregma to lambda	Martin 1928
Arc from lambda to opisthion	Martin 1928
Minimum width of the ramus	Martin 1928
Maximum breadth between the mandibular condyles	Martin 1928
Maximum breadth between the mandibular angles	Martin 1928
Maximum breadth between the mental foramina	Martin 1928
Height of the mandibular symphysis in the mid-sagittal plane	Martin 1928
Maximum width of the mandibular body at the level of the mental foramen	Martin 1928

REFERENCES

Alekseev, V. P., and G. F. Debets. 1964. *Kraniometria. Metodika anthropologitsheskh isledovaniy.* Izd. Nauka, Moskva.

Evteev, A. A., A. A. Movsesian, and A. N. Grosheva. 2017. The association between midfacial morphology and climate in northeast Europe differs from that in north Asia: Implications for understanding the morphology of Late Pleistocene *Homo sapiens. Journal of Human Evolution* 107: 36–48.

Howells, W. W. 1989. *Skull shapes and the map.* Cambridge: Peabody Museum Press.

Martin, R. 1928. *Lehrbuch der Anthropologie in Systematischer darstellung.* 2-e Bd. *Kraniologie. Osteologie.* Jena.

Table S2.
List of genomic populations and sample sizes for Affymetrix dataset (50786 SNPs), paired with cranial populations.

Genetic Population (Affymetrix)	N	References	Cranial Population Match
Abkhasian	9	Lazaridis et al. 2014	Abkhazian
Adygei	17	Lazaridis et al. 2014	Shapsug
Armenian	10	Lazaridis et al. 2014	Armenian
Armenian_WGA	3	Lazaridis et al. 2014	Armenian
Bulgaria_POPRES	2	Nelson et al. 2008	Bulgarian
Bulgarian	10	Lazaridis et al. 2014	Bulgarian
Buryat	25	Xing et al. 2010	Buryat
CHB(Han_Chinese_in_Beijing_China)	84	Altschuler et al. 2010	Han (North China)
Han_NChina	10	Lazaridis et al. 2014	Han (North China)
Chukchi	20	Lazaridis et al. 2014	Chukchi
Druze	39	Lazaridis et al. 2014	Druze
Eskimo_Naukan	13	Lazaridis et al. 2014	Eskimo (Siberian Yupik)
Finland_POPRES	2	Nelson et al. 2008	Finn
Finnish	104	Auton et al. 2015	Finn
TSI(Tuscans_italy)	88	Altschuler et al. 2010	Italian
Italian_Bergamo	12	Lazaridis et al. 2014	Italian
Italian_South	1	Lazaridis et al. 2014	Italian
Italian_Tuscan	8	Lazaridis et al. 2014	Italian
Italy	213	Nelson et al. 2008	Italian
Japan_POPRES	68	Nelson et al. 2008	Japanese
Japanese	29	Lazaridis et al. 2014	Japanese
Japanese	13	Xing et al. 2009	Japanese
JPT(Japanese_in_Tokio)	91	Altschuler et al. 2010	Japanese
Mansi	8	Lazaridis et al. 2014	Mansi
Mongola	6	Lazaridis et al. 2014	Mongol
Mongola	2	López-Herráez et al. 2009	Mongol
Mordovian	10	Lazaridis et al. 2014	Mordovian
North_Ossetian	10	Lazaridis et al. 2014	Ossetian
Norway_POPRES	2	Nelson et al. 2008	Norse
Norwegian	11	Lazaridis et al. 2014	Norse
Romania_POPRES	16	Nelson et al. 2008	Romanian

cont. ⟶

Genetic Population (Affymetrix)	N	References	Cranial Population Match
Russia_POPRES	8	Nelson et al. 2008	Russian
Russian	22	Lazaridis et al. 2014	Russian
Russian	1	López-Herráez et al. 2009	Russian
Turkey_POPRES	7	Nelson et al. 2008	Turk
Turkish	4	Lazaridis et al. 2014	Turk
Turkish_Adana	10	Lazaridis et al. 2014	Turk
Turkish_Aydin	7	Lazaridis et al. 2014	Turk
Turkish_Balikesir	6	Lazaridis et al. 2014	Turk
Turkish_Istanbul	10	Lazaridis et al. 2014	Turk
Turkish_Kayseri	10	Lazaridis et al. 2014	Turk
Turkish_Trabzon	9	Lazaridis et al. 2014	Turk
Tuvinian	10	Lazaridis et al. 2014	Tuvinian
Ukraine_POPRES	2	Nelson et al. 2008	Ukrainian
Ukrainian_East	6	Lazaridis et al. 2014	Ukrainian
Ulchi	25	Lazaridis et al. 2014	Ulchi
Yakut	20	Lazaridis et al. 2014	Yakut
Total number of individuals:	**1093**		

REFERENCES

Altshuler, D. M., R. A. Gibbs, L. Peltonen, E. Dermitzakis, S. F. Schaffner, F. Yu, P. E. Bonnen, P. I. W. de Bakker, et al. 2010. Integrating common and rare genetic variation in diverse human populations. *Nature* 467 (7311): 52–58.

Auton, A., G. R. Abecasis, D. M. Altshuler, R. M. Durbin, D. R. Bentley, A. Chakravarti, et al. 2015. A global reference for human genetic variation. *Nature* 526 (7571): 68–74.

Nelson, M. R., K. Bryc, K. S. King, A. Indap, A. R. Boyko, J. Novembre, et al. 2008. The population reference sample, POPRES: A resource for population, disease, and pharmacological genetics research. *The American Journal of Human Genetics* 83 (3): 347–358.

Lazaridis, I., N. Patterson, A. Mittnik, G. Renaud, S. Mallick, K. Kirsanow, P. H. Sudmant, J. G. Schraiber, S. Castellano, M. Lipson, B. Berger, C. Economou, R. Bollongino, et al. 2014. Ancient human genomes suggest three ancestral populations for present-day Europeans. *Nature* 513 (7518): 409–413.

López Herráez, D., M. Bauchet, K. Tang, C. Theunert, I. Pugach, J. Li, M.R. Nandineni, A. Gross, M. Scholz, and M. Stoneking. 2009. Genetic variation and recent positive selection in worldwide human populations: Evidence from nearly 1 million SNPs. *PLOS ONE* 4(11): e7888.

Xing, J., W. S. Watkins, D. J. Witherspoon, Y. Zhang, S. L. Guthery, R. Thara, B. J. Mowry, K. Bulayeva, R. B. Weiss, and L. B. Jorde. 2009. Fine-scaled human genetic structure revealed by SNP microarrays. *Genome Research* 19 (5): 815–825.

Xing, Jinchuan, W. Scott Watkins, Adam Shlien, Erin Walker, Chad D. Huff, David J. Witherspoon, Yuhua Zhang, Tatum S. Simonson, Robert B. Weiss, Joshua D. Schiffman, David Malkin, Scott R. Woodward, and Lynn B. Jorde. 2010. Toward a more uniform sampling of human genetic diversity: A survey of worldwide populations by high-density genotyping. *Genomics* 96 (4): 199–210.

Table S3.
List of genomic populations and sample sizes for Illumina dataset (114109 SNPs), paired with cranial populations.

Genetic Population (Illumina)	N	References	Cranial Population Match
Finland	157	Leu et al. 2010	Finn
Finnish	100	Auton et al. 2015	Finn
Finnish	2	Hellenthal et al. 2014	Finn
Karelians	15	Yunusbayev et al. 2015	Karelian
Komis	16	Yunusbayev et al. 2015	Komi
Mordovian	15	Yunusbayev et al. 2012	Mordovian
Russian_central and South	32	Yunusbayev et al. 2015	Russian
Ukranian	20	Yunusbayev et al. 2012	Ukrainian
Total number of individuals:	**357**		

REFERENCES

Auton, A., G. R. Abecasis, D. M. Altshuler, R. M. Durbin, D. R. Bentley, A. Chakravarti, et al. 2015. A global reference for human genetic variation. *Nature* 526(7571): 68–74.

Hellenthal, G., G. B. J. Busby, G. Band, J. F. Wilson, C. Capelli, D. Falush, and S. Myers. 2014. A genetic atlas of human admixture history. *Science* 343 (6172): 747–751.

Leu, M., K. Humphreys, I. Surakka, E. Rehnberg, J. Muilu, P. Rosenström, P. Almgren, J. Jääskeläinen, R. P. Lifton, K. Ohm Kyvik, J. Kaprio, N. L. Pedersen, A. Palotie, P. Hall, H. Grönberg, L. Groop, L. Peltonen, J. Palmgren, and S. Ripatti. 2010. NordicDB: a Nordic pool and portal for genome-wide control data. *European Journal of Human Genetics* 18 (12): 1322–1326.

Yunusbayev, B., M.Metspalu, M. Järve, I. Kutuev, S. Rootsi, E. Metspalu, D. M. Behar, K.Varendi, H. Sahakyan, R. Khusainova, L. Yepiskoposyan, E. K. Khusnutdinova, P. A. Underhill, T. Kivisild, and R. Villems. 2012. The Caucasus as an asymmetric semipermeable barrier to ancient human migrations. *Molecular Biology and Evolution* 29 (1): 359–365.

Yunusbayev, B., M. Metspalu, E. Metspalu, A. Valeev, S. Litvinov, R. Valiev, V. Akhmetova, E. Balanovska, O. Balanovsky, S. Turdikulova, D. Dalimova, P. Nymadawa, A. Bahmanimehr, H. Sahakyan, K. Tambets, S. Fedorova, N. Barashkov, I. Khidiyatova, E. Mihailov, R. Khusainova, L. Damba, M. Derenko, B. Malyarchuk, L. Osipova, M. Voevoda, L. Yepiskoposyan, T. Kivisild, E. Khusnutdinova, and R. Villems. 2015. The genetic legacy of the expansion of Turkic-speaking nomads across Eurasia. *PLOS Genetics* 11 (4):e1005068.

Table S4.
Sample sizes for paired uni-parental genetic data (mitochondrial and Y-Chromosome DNA polymorphisms).

Cranial Population	mtDNA*	Y-chromosome**
Abkhazian	137 (Kutuev 2010)	162 (Kutuev 2010)
Armenian	191 (Richards 2000)	57 (Kutuev 2010)
Bolgarian	855 (Karachanak 2012)	808 (Karachanak 2013)
Buryat	101 (Pakendorf et al. 2003)	81 (Jin et al. 2009)
Chukchi	24 (Starikovskaya et al. 1998)	No data
Druze	622 (Shlush 2008)	109 (Zalloua 2008)
Eskimo (Siberian Yupik)	50 (Raff et al. 2015)	33 (Lell et al. 2002)
Evenk	40 (Pakendorf et al. 2003)	96 (Tambets et al. 2004)
Finn	432 (Meinila 2001; Richards 1996)	316 (Lappalainen 2006; Heinrich 2009)
Han (North Chinese)	322 (Jin et al. 2009)	242 (Jin et al. 2009)
Italian	70 (Rienzo 1991)	347 (Coia 2013; Batini 2015)
Japanese	211 (Jin et al. 2009)	154 (Jin et al. 2009)
Karelian	303 (Lappalainen 2008)	202 (Lappalainen 2006, 2008)
Khanty	209 (Pimenoff et al. 2006)	27 (Pimenoff et al. 2006)
Komi	214 (Osipova 2005)	153 (Mirabal 2009; Trofimova 2015)
Latvian	299 (Pliss 2006)	159 (Pliss 2015)
Mansi	95 (Pimenoff et al. 2006)	25 (Pimenoff et al. 2006)
Mongol	95 (Jin et al. 2009)	65 (Jin et al. 2009)
Mordovian	102 (Bermisheva 2002)	59 (Trofimova 2015)
Norse	74 (Passarino 2002)	1789 (Dupuy 2006; Batini 2015)
Ossetian	162 (Kutuev 2010)	153 (Kutuev 2010)
Romanian	146 (Jankova-Ajanovska 2014)	67 (Bosch 2005)
Russian	306 (Malyarchuk 2004)	183 (Fechner 2008; Malyarchuk 2008; Mirabal 2009)
Saami	637 (Ingman 2007)	189 (Dupuy 2006; Batini 2015)
Shapsug (Adygean)	155 (Kutuev 2010)	154 (Kutuev 2010)
Turk	190 (Quintana-Murci 2004; Jankova-Ajanovska 2014)	20 (Batini 2015)
Ukrainian	680 (Pshenichnov 2013)	154 (Mielnik-Sikorska 2013)
Ulchi	160 (Sukernik et al. 2012)	53 (Lell et al. 2002) (combined with the Nanai)
Yakut	83 (Pakendorf et al. 2003)	155 (Tambets et al. 2004)
Total	**6965**	**6012**

* Polymorphisms: A, B, CZ, D, F, G, HV, H, I, J, K, L, M, N, R, T, U*, U1, U2, U3, U4, U5, U6, U7, U8, V, W, X, Y.
** Polymorphisms: B, C, D, E, F, G, H, I, IJ, J, K, L, N, P, Q, R1a, R1b, R*, T.

REFERENCES

Batini, C., P. Hallas, D. Zadik, P. M. Delser, A. Benazzo, S. Ghirotto, E. Arroyo-Pardo, G. L. Cavalleri, P. de Knijff, B. M. Dupuy, H. A. Eriksen, T. E. King, A. L. de Munain, A. M. López-Parra, A. Loutradis, J. Milasin, A. Novelletto, H. Pamjay, A. Sajantila, A. Tolun, B. Winney, and M. A. Jobling. 2015. Large-scale recent expansion of European patrilineages shown by population resequencing. *Nature Communications* 6: 7152.

Bermisheva, M., K. Tambets, R. Villems, and A. E. Khusnutdinova. 2002. Diversity of mitochondrial DNA haplotypes in ethnic populations of the Volga-Ural region of Russia. *Molecular Biology* (Russian) 36(6): 990–1001.

Bosch, E., F. Calafell, A. González-Neira, C. Flaiz, E. Mateu, H.-G. Scheil, W. Huckenbeck, L. Efremovska, I. Mikerezi, N. Xirotiris, C. Grasa, H. Schmidt, and D. Comas. 2006. Paternal and maternal lineages in the Balkans show a homogenous landscape over linguistic barriers, except for the isolated Aromuns. *Annals of Human Genetics* 70: 459–487.

Coia, V., M. Capocasa, P. Anagnostou, V. Pascali, F. Scarnicci, I. Boschi, C. Battaggia, F. Crivellaro, G. Ferri, M. Alù, F. Brisighelli, G. B. J. Busby, C. Capelli, F. Maixner, G. Cipollini, P. P. Viazzo, A. Zink, and G. Destro Bisol. 2013. Demographic histories, isolation and social factors as determinants of the genetic structure of alpine linguistic groups. *PLoS ONE* 8(12): e81704. doi:10.1371/journal.pone.008170

Crawford, M., R. C. Rubicz, and M. Zlojutro. 2010. Origins of Aleuts and the genetic structure of populations of the archipelago: Molecular and archaeological perspectives. *Human Biology* 82(5-6): 695–717.

Dupuy B. M., M. Stenersen, T. T. Lu, and B. Olaisen. 2006. Geographical heterogeneity of Y-chromosomal lineages in Norway. *Forensic Science International* 164(1):10-9

Fechner, A., D. Quinque, S. Rychov, I. Morozowa, O. Naumova, Y. Schnieder, S. Willuweit, O. Zhukova, L. Roewer, M. Stoneking, and I. Nasidze. 2008. Boundaries and clines of the west Eurasian Y-chromosome landscape: Insights from the European part of Russia. *American Journal of Physical Anthropology* 137: 41–47.

Heinrich, M., T. Braun, T. Sanger, P. Saukko, S. Lutz-Bonengel, and U. Schmidt. 2009. Reduced-volume and low-volume typing of Y-chromoisomal SNPs to obtain Finnish Y-chromosomal compound haplotypes. *International Journal of Legal Medicine* 123: 413–418.

Ingman, M., and U. Gyllensten. 2007. A recent genetic link between Sami and the Volga-Ural region of Russia Europ. *Journal of Human Genetics* 15: 115–120.

Jankova-Ajanovska, R., B. Zimmermann, G. Huber, A. W. Röck, M. Bodner, Z. Jakovski, B. Janeska, A. Duma, and W. Parson. 2014. Mitochondrial DNA control region analysis of three ethnic groups in the Republic of Macedonia. *Genetics* 13: 1–2.

Jin, H-J, C. Tyler-Smith, and W. Kim. 2009. The peopling of Korea revealed by analyses of mitochondrial DNA and Y-chromosomal markers. *PLoS ONE* 4(1): e4210. https://doi.org/10.1371/journal.pone.0004210

Karachanak, S., V. Carossa, D. Nesheva, A. Olivieri, M. Pala, B. H. Kashani, V. Grugni, V. Battaglia, A. Achilli, Y. Yordano, A. S. Galabov, O. Semino, D. Toncheva, and A. Torroni. 2011. Bulgarians vs the other European populations: A mitochondrial DNA perspective. *International Journal of Legal Medicine*. doi: 10.1007/s00414-011-0589-y

Kutuev, I. A. 2010. Geneticheskaya struktura i philogeografiya narodov Kavkaza (Genetic structure and phylogeography of peoples of the Caucasus). Doctor of Science Dissertation, Ufa. Institute of biochemistry and genetics of the Ufa scientific centre of Russian Academy of Science.

Lappalainen, T., S. Koivumaki, E. Salmela, K. Huoponen, P. Sistonen, M.-L. Savontaus, and P. Lahermo. 2006. Regional differences among the Finns: A Y-chromosomal perspective. *Gene* 376: 207–215.

Lappalainen, T., V. Laitinen, E. Salmela, P. Andersen, K. Huoponen, M.-L. Savontaus, and P. Lahermo. 2008. Migration waves to the Baltic Sea region. *Annals of Human Genetics* 72: 337–348.

Lell J. T., R. I. Sukernik, Y. B. Starikovskaya, B. Su, L. Jin, T. G. Schurr, P. A. Underhill, and D. C. Wallace. 2002. The dual origin and Siberian affinities of Native American Y chromosomes. *American Journal of Human Genetics* 70(1): 192–206. doi: 10.1086/338457

Malyarchuk, B. A., M. V. Derenko, T. Grzybowski, A. Lunkina, J. Czarny, S. Rychkov, I. Morozova, G. Denisova, and D. Miscicka-Sliwka. 2004. Differentiation of mitochondrial DNA and Y-chromosomes in Russian populations. *Human Biology* 76(6): 877–900.

Malyarchuk, B. A., and M. V. Derenko. 2008. Gene pool structure of Russian populations from the European part of Russia inferred from the data on Y-chromosome haplogroups distribution. *Russian Journal of Genetics* 44 (2C): 187–192.

Meinila, M., S. Finnila, and K. Majamaa. 2001. Evidence for mtDNA admixture between the Finns and the Saami. *Human Heredity* 52(3): 160–170.

Mielnik-Sikorska, M., P. Daca, M. Wozniak, B. A. Malyarchuk, J. Bednarek, T. Dobosz, and T. Grzybowski. 2013. Genetic data from Y-chromosome STR and SNP loci in Ukrainian population. *Genetics* 7: 200–203.

Mirabal, S., M. Regueiro, A. M. Cadenas, L. L. Cavalli-Sforza, P. A. Underhill, D. A. Verbenko, S. A. Limborska, and R. J. Herrera. 2009. Y-chromosome distribution within the geo-linguistic landscape of northwestern Russia. *European Journal of Human Genetics* 17: 1260–1273.

Morozova, I., A. Evsyukov, A. Kon'kov, A. Grosheva, O. Zhukova, and S. Rychov. 2012. Russian ethnic history inferred from mitochondrial DNA diversity. *American Journal of Physical Anthropology* 147: 341–351.

Osipova, L.P. 2005. Korennoe naselenie Shuryshkarskogo rayona Yamalo-Nenetskogo avtonomnogo okruga: demographicheskie, geneticheskie i meditsinskie aspekty (Indigenous population of Shuryshkarsky district of Yanalo-Nenetsky autonomous okrug: demographic, genetic and medical aspects). ART–AVENUE, Novosibirsk.

Pakendorf, B., V. Wiebe, L. A. Tarskaia, V. A. Spitsyn, H. Soodyall, A. Rodewald, and M. Stoneking. 2003. Mitochondrial DNA evidence for admixed origins of central Siberian populations. *American Journal of Physical Anthropology* 120(3): 211–24.

Passarino, G., G. L. Cavalleri, A. A. Lin, L. L. Cavalli-Sforza, A.-L. Børresen-Dale, and P. A. Underhill. 2002. Different genetic components in the Norwegian population revealed by the analysis of mtDNA and Y-chromosome polymorphisms. *European Journal of Human Genetics* 10: 521–529.

Pimenoff,V.N., D. Comas, J. U. Palo, G. Vershubsky, A. Kozlov, and A. Sajantila. 2008. Northwest Siberian Khanty and Mansi in the junction of West and East Eurasian gene pools as revealed by uniparental markers. *European Journal of Human Genetics* 16: 1254–1264.

Pliss, L., L. Timša, S. Rootsi, K. Tambets, I. Pelnena, E. Zole, A. Puzuka, A. Sabule, S. Rozane, B. Lace, V. Kucinskas, A. Krumina, R. Ranka, and V. Baumanis. 2015. Y-chromosomal lineages of Latvians in the context of the genetic variation of the eastern Baltic Region. *Annals of Human Genetics* 79: 418–430.

Pshenichnov, A., O. Balanovsky, O. Utevska, E. Metspalu, V. Zaporozhchenko, A. Agdzhoyan, M. Churnosov, L. Atramentova, and E. Balanovska. 2013 Genetic affinities of Ukrainians from the maternal perspective. *American Journal of Physical Anthropology* 152(4): 543–550.

Quintana-Murci, L., R. Chaix, R. S. Wells, D. M. Behar, H. Sayar, R. Scozzari, C. Rengo, N. Al-Zahery, O. Semino, A. S. Santachiara-Benerecetti, A. Coppa, Q. Ayub, A. Mohyuddin, C. Tyler-Smith, S. Q. Mehdi, A. Torroni, and K. McElreavey. 2004. Where west meets east: The complex mtDNA landscape of the southwest and Central Asian corridor. *American Journal of Human Genetics* 74(5): 827–845.

Raff, J.A., M. Rzhetskaya, J. Tackney, and M. G. Hayes. 2015. Mitochondrial diversity of Iñupiat people from the Alaskan North Slope provides evidence for the origins of the Paleo- and Neo-Eskimo peoples. *American Journal of Physical Anthropology* 157(4): 603–14. doi: 10.1002/ajpa.22750. Epub 2015 Apr 17.

Richards, M., H. Côrte-Real, P. Forster, V. Macaulay, H. Wilkinson-Herbots, A. Demaine, S. Papiha, R. Hedges, H. J. Bandelt, and B. Sykes. 1996. Paleolithic and neolithic lineages in the European mitochondrial gene pool. *American Journal of Human Genetics* 59(1): 185–203.

Richards, M., V. Macaulay, E. Hickey, E. Vega, B. Sykes, V. Guida, C. Rengo, D. Sellitto, F. Cruciani, et al. 2000. Tracing European founder lineages in the Near Eastern mtDNA pool. *American Journal of Human Genetics* 67(5): 1251–1276.

Rienzo, A. D., and A. C. Willson. 1991. Branching pattern in the evolutionary tree for human mitochondrial DNA. *Proceedings of the National Academy of Science USA* 88(5): 1597–1601.

Rubicz, R., M. Zlojutro, G. Sun, V. Spitsyn, R. Deka, K. L. Young, and M. H. Crawford. 2010. Genetic architecture of a small, recently aggregated Aleut population: Bering Island, Russia. *Human Biology* 82(5/6): 719-36. https://doi.org/10.3378/027.082.0512

Shlush, L. I., D. M. Behar, G. Yudkovsky, A. Templeton, Y. Hadid, F. Basis, M. Hammer, S. Itzkovitz, and K. Skorecki. 2008. The Druze: A population genetic refugium of the Near East. *PLoS One* 3(5): e2105.

Starikovskaya, Y.B., R. I. Sukernik, T. G. Schurr, A. M. Kogelnik, and D. C. Wallace. 1998. mtDNA diversity in Chukchi and Siberian Eskimos: implications for the genetic history of Ancient Beringia and the peopling of the New World. *American Journal of Human Genetics* 63(5): 1473-91. doi: 10.1086/302087. PubMed PMID: 9792876

Sukernik, R. I., N. V. Volodko, I. O. Mazunin, N. P. Eltsov, S. V. Dryomov, and E. B. Starikovskaya. 2012. Mitochondrial genome diversity in the Tubalar, Even, and Ulchi: Contribution to prehistory of native Siberians and their affinities to Native Americans. *American Journal of Physical Anthropology* 148(1): 123–38. doi: 10.1002/ajpa.22050. Epub 2012 Apr 4.

Tambets, K., S. Rootsi, T. Kivisild, H. Help, P. Serk, E. L. Loogväli, H. V. Tolk, et al. 2004. The western and eastern roots of the Saami--the story of genetic "outliers" told by mitochondrial DNA and Y chromosomes. *American Journal of Human Genetics* 74(4): 661–82. doi: 10.1086/383203

Trofimova, N.V. 2015. Izmenchivost' mitokhondrial'noy DNK and Y-khromosomy v populiatsiyakh volgo-uralskogo regiona (The mtDNA and Y-chromosome diversity in populations of Volga-Ural region). Ph.D. Dissertation, Ufa. Institute of Biochemistry and Genetics of the Ufa Scientific Centre of Russian Academy of Sciences. [In Russian]

Zalloua, P. A., Y. Xue, J. Khalife, N. Makhoul, L. Debiane, D. E. Platt, A. K. Royyuru, R. J. Herrera, D. F. S. Hernanz, J. Blue-Smith, R. S. Wells, D. Comas, J. Bertranpetit, C. Tyler-Smith, The Genographic Consortium. 2008. Y-chromosomal diversity in Lebanon is structured by recent historical events. *American Journal of Human Genetics* 82: 873–882.

Table S5.
Correlations between the matrices of craniometric, genetic, and geographic distances for size-standardized craniometric variables.

Geographical Scale		SNP		mtDNA		Y-Chromosme		Geography	
		r-value	p-value	r-value	p-value	r-value	p-value	r-value	p-value
Eurasia	mid-face	0.904	0.001***	0.789	0.001***	0.589	0.001***	0.800	0.001***
	vault	0.409	0.001***	0.427	0.001***	0.441	0.001***	0.388	0.001***
	cranium	0.805	0.001***	0.700	0.001***	0.570	0.001***	0.693	0.001***
	mandible	0.243	0.031*	0.173	0.105	0.247	0.007**	0.279	0.011*
	mean	0.590		0.522		0.462		0.540	
East Eurasia	mid-face	0.639	0.011*	0.446	0.014*	0.463	0.004**	0.511	0.002**
	vault	0.152	0.377	0.275	0.085	0.413	0.016*	0.421	0.01**
	cranium	0.438	0.028*	0.356	0.050	0.553	0.003**	0.496	0.007**
	mandible	0.141	0.535	0.057	0.793	0.332	0.063	0.145	0.427
	mean	0.342		0.284		0.440		0.393	
West Eurasia	mid-face	0.631	0.001***	0.170	0.262	0.393	0.003**	0.265	0.017*
	vault	0.169	0.280	0.088	0.622	0.028	0.801	0.210	0.092
	cranium	0.318	0.04*	0.123	0.509	0.196	0.104	0.271	0.042*
	mandible	0.849	0.002**	0.165	0.598	0.395	0.048*	0.283	0.031*
	mean	0.492		0.136		0.253		0.257	
North Asia	mid-face	0.765	0.002**	0.676	0.001***	0.222	0.259	0.537	0.001***
	vault	0.207	0.272	0.281	0.098	-0.173	0.385	0.460	0.009**
	cranium	0.576	0.01**	0.507	0.006**	-0.015	0.935	0.587	0.002**
	mandible	-0.245	0.477	-0.146	0.564	-0.053	0.812	-0.075	0.678
	mean	0.326		0.330		-0.005		0.377	
Europe-Mediterranean	mid-face	0.321	0.105	0.085	0.682	0.091	0.638	0.122	0.538
	vault	0.188	0.447	0.081	0.713	-0.033	0.885	0.039	0.887
	cranium	0.221	0.363	0.116	0.630	0.060	0.730	0.121	0.604
	mean	0.264		0.153		0.028		0.165	
Northeast Europe	mid-face	0.139	0.687	-0.117	0.741	0.490	0.055	0.242	0.330
	vault	0.190	0.699	0.790	0.065	0.213	0.417	0.182	0.447
	cranium	0.153	0.695	0.537	0.208	0.292	0.260	0.193	0.456
	mandible	0.002	1.000	0.850	0.049*	0.125	0.786	0.012	0.962
	mean	0.121		0.515		0.280		0.157	

Notes: r-values are Pearson correlation coefficients; statistical significance after 1000 permutations: two-tailed p-value *<0.05; **<0.01; ***<0.001.